REVISE EDEXCEL GCSE

Science

Additional Science

REVISION GUIDE

Higher

Series Consultant: Harry Smith
Series Editor: Penny Johnson

Authors: Penny Johnson, Sue Kearsey, Damian Riddle

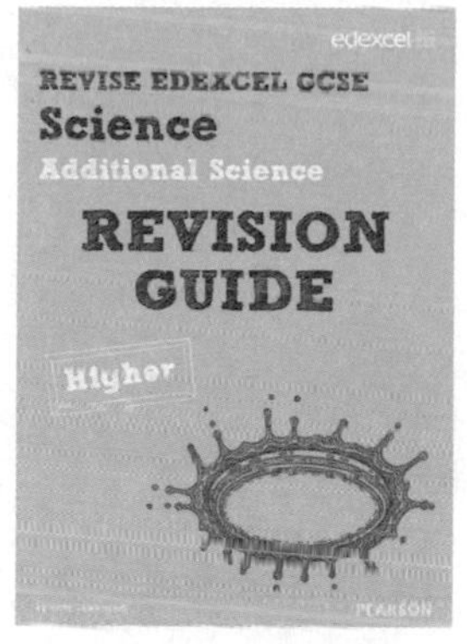

THE REVISE EDEXCEL SERIES

Available in print or online

Online editions for all titles in the Revise Edexcel series are available Autumn 2012.

Presented on our ActiveLearn platform, you can view the full book and customise it by adding notes, comments and weblinks.

Print editions

Additional Science Revision Guide Higher	9781446902653
Additional Science Revision Workbook Higher	9781446902660

Online editions

Additional Science Revision Guide Higher	9781446904596
Additional Science Revision Workbook Higher	9781446904602

Print and online editions are also available for Science (Higher and Foundation), Additional Science (Foundation) and Extension Units.

This Revision Guide is designed to complement your classroom and home learning, and to help prepare you for the exam. It does not include all the content and skills needed for the complete course. It is designed to work in combination with Edexcel's main GCSE Science 2011 Series.

800

To find out more visit:
www.pears[...]celgcsesciencerevision

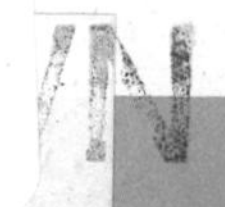

ALWAYS LEARNING

PEARSON

Contents

BIOLOGY

CHEMISTRY

PHYSICS

A small bit of small print

Target grade ranges are quoted in this book for some of the questions. Students targeting this grade range should be aiming to get most of the marks available. Students targeting a higher grade should be aiming to get all of the marks available.

Edexcel publishes Sample Assessment Material and the Specification on its website. This is the official content and this book should be used in conjunction with it. The questions in Now try this have been written to help you practise every topic in the book. Remember: the real exam questions may not look like this.

Plant and animal cells

Animals and plants are formed from cells. Animal cells and plant cells have some parts in common. These parts have particular functions in a cell.

animal cell

cell membrane controls what enters and leaves the cell, e.g. oxygen, carbon dioxide, glucose

nucleus a large structure that contains DNA – instructions for the building and working of the cell

cytoplasm jelly-like substance that fills the cell – many reactions take place here

mitochondria (single: mitochondrion) – tiny structures where respiration takes place, releasing energy for cell processes

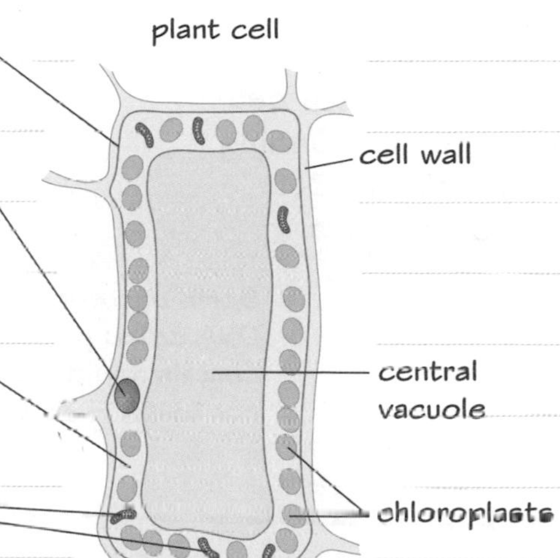

The cytoplasm of a cell is not a proper structure, such as a membrane or chloroplast. It is the jelly-like substance that fills the cell and supports other structures.

Worked example

Name the three structures that are found in most plant cells but not animal cells, and describe their functions.

Chloroplasts are the structures where photosynthesis takes place to make food for the plant cell.

The cell wall is made of cellulose, and is tough so that it helps support the cell and helps it keep its shape.

The large central vacuole contains cell sap, which helps to keep the plant cell rigid.

Watch out! The cell membrane and cell wall are different and separate structures.

Now try this

target D-C

1. Draw up a table like the one below to show the main components of plant and animal cells and their functions.

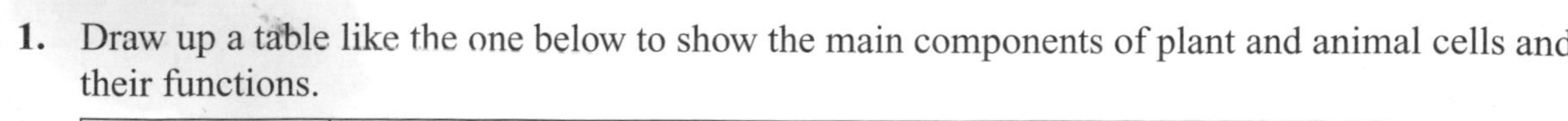

Component	Found in plant cells?	Found in animal cells?	Function

(7 marks)

target D-B

2. Plants don't have skeletons. Explain how they stand upright. (2 marks)

target C-A

3. Explain why not all plant cells have chloroplasts. (2 marks)

Inside bacteria

Bacterial cells

Bacteria have a simple cell structure. Like animal and plant cells, they have a cell membrane surrounding the cytoplasm. But they do not have a nucleus.

A single loop of chromosomal DNA lies free in the cytoplasm. This carries most of the bacterial genes.
cell membrane
Some bacteria have a flagellum to help them move.
Some bacteria have extra circles of DNA called plasmid DNA. Plasmids contain additional genes that are not found in chromosomes.
Many bacteria have a cell wall for protection, but it is made of different substances to plant cell walls.

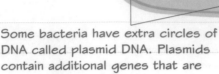

Using microscopes

Before microscopes were invented about 350 years ago, we could not see the cells in organisms. Magnification enables us to see plant cells, animal cells and bacterial cells, and the structures inside them.

A light microscope uses light to magnify objects. The greatest possible magnification using a light microscope is about ×2000.

An electron microscope uses electrons to view an object. This makes it possible to magnify objects up to about ×10 million. We can see far more detail in cells with an electron microscope than with a light microscope.

Worked example

Some cells were viewed by microscope using a ×4 eyepiece and ×20 objective. Calculate the magnification of the cells seen through the microscope.

magnification of object

= magnification of eyepiece × magnification of objective

$= 4 \times 20 = 80$

The cells will be magnified 80 times by the microscope.

EXAM ALERT!

Always show your working in a calculation. Even if you get the answer wrong you may be able to show that you understand the method.

Students have struggled with exam questions similar to this - **be prepared!** ResultsPlus

Now try this

target
D-B

1. A bacterium is viewed under a light microscope using a ×40 objective and a ×10 eyepiece. The image is 1.2 mm long. Calculate the actual length of the cell. **(2 marks)**

2. Compare the structure and function of the two types of DNA in bacteria. **(2 marks)**

DNA

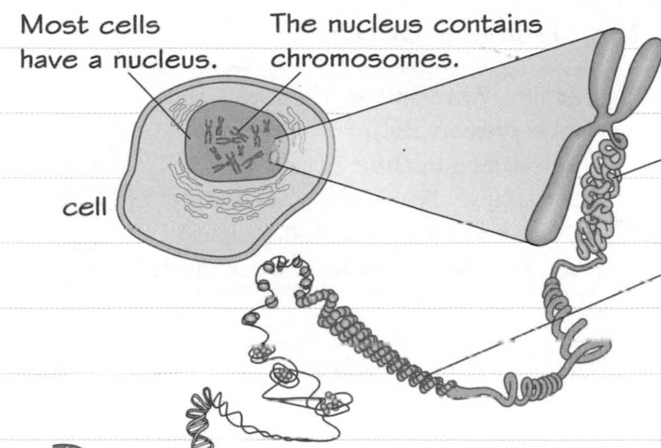

Most cells have a nucleus.
The nucleus contains chromosomes.
cell

chromosome

A chromosome consists of a string of genes.

A gene is a short piece of DNA that codes for a specific protein. You have genes for hair structure, eye colour, enzymes and every other protein in your body.

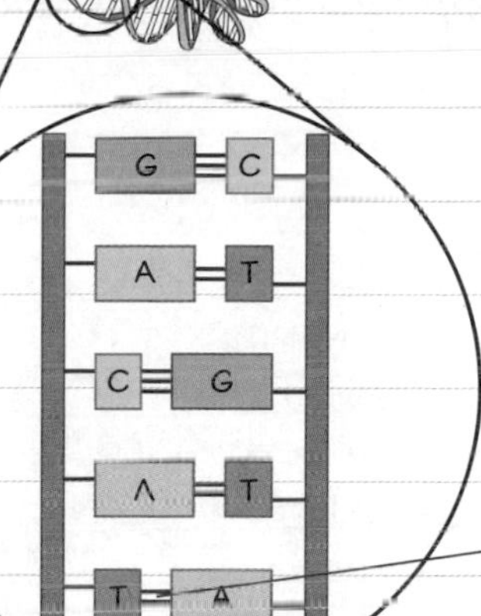

Each gene is a length of DNA. DNA is a long, coiled molecule formed from two strands. The strands are twisted in a double helix.

The two strands of the double helix are joined by pairs of bases. There are four different bases in DNA:
A = adenine T = thymine
C = cytosine G = guanine

Bases form complementary pairs:
A always pairs with T
C always pairs with G.

Remember: straight A with straight T; curly C with curly G.

Weak hydrogen bonds between the base pairs hold the DNA strands together.

Extracting DNA

DNA can be extracted from plant tissue, such as from kiwi fruit.

Kiwi fruit is mashed up and mixed with salty water and detergent. This breaks open the cells and helps to release the DNA from the nuclei. → Protease enzyme is added to the filtered mixture. The enzyme helps to break up proteins in cell membranes and so release more DNA. → Ice-cold ethanol is poured carefully down inside the tube into the mixture. The ethanol makes the DNA separate from the liquid so it is easy to lift out.

Now try this

target D-C

1. Write a sentence to define each of these words:
   (a) gene (b) base. **(2 marks)**

target D-B

2. The sequence of bases on one strand of DNA is CGAT. Write down the sequence of bases on the complementary strand, and explain how you worked out your answer. **(2 marks)**

target C-A

3. Describe the structure of DNA. **(4 marks)**

This question has 4 marks, so the answer needs 4 different ideas.

DNA discovery

Several scientists played a key role in working out the structure of DNA.

Maurice Wilkins and **Rosalind Franklin** studied the structure of DNA using X-rays.

Franklin studied X-ray photographs of DNA to work out how the atoms were grouped.

James Watson and **Francis Crick** interpreted data from other scientists in their study of DNA structure. Franklin's photographs were the final clue that helped them build the double helix model.

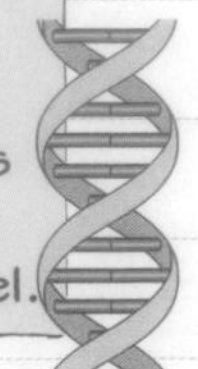

The Human Genome Project

Scientists in 18 different countries collaborated to decode the human genome. This is the order of bases on all the human chromosomes. The project was completed quickly because so many different scientists worked on it at the same time. The work was published in 2003 so that any scientist can use the information to develop new medicines and techniques.

Worked example

Describe two possible developments as a result of decoding the human genome, and discuss the implications of these developments.

One development is the identification of genes that can cause disease. Knowing if you have a faulty gene could help a person and their family prepare for its effects, but some people would prefer not to know if they have a faulty gene because then they would worry about it.

Another development is gene therapy. This involves replacing faulty alleles in body cells with healthy ones. This would allow the affected person to live a normal life. However, people will have to decide whether the faulty alleles are replaced in gametes, which would mean the alleles could be passed on to children.

There are many possible answers for this question, because there are many new developments. Other possibilities include: personalised medicines, and the evolutionary relationships between humans and other organisms. As well as learning about new developments you need to be able to say what the implications are.

Now try this

1. Suggest why Watson, Crick and Wilkins were all given the Nobel Prize for the discovery of the structure of DNA. **(1 mark)**
2. Use the examples of the Human Genome Project and the discovery of DNA to describe some advantages of collaboration between scientists. **(2 marks)**

Genetic engineering

In genetic engineering (also called genetic modification) a gene from one organism is inserted into the DNA of another organism. The inserted gene then makes its protein in the genetically modified (GM) organism.

1. A gene (e.g. for insulin) is cut out of a human chromosome using enzymes.

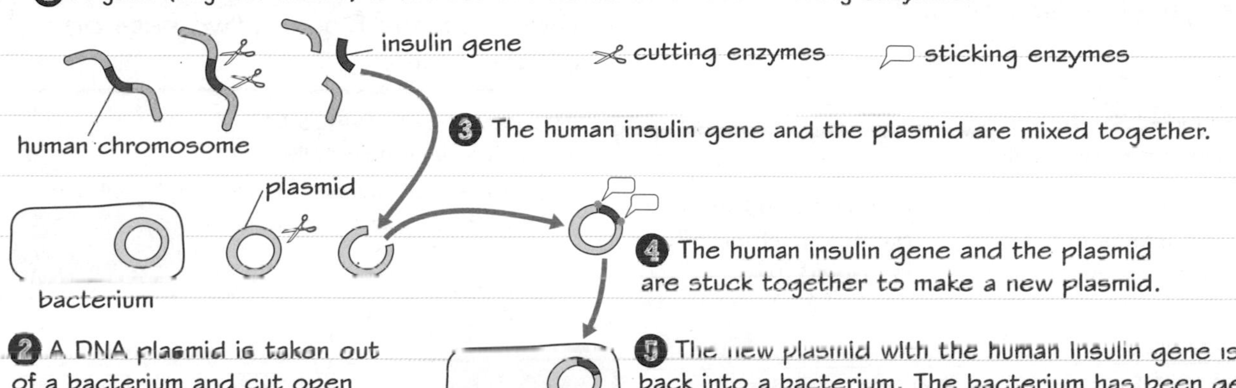

2. A DNA plasmid is taken out of a bacterium and cut open using enzymes.
3. The human insulin gene and the plasmid are mixed together.
4. The human insulin gene and the plasmid are stuck together to make a new plasmid.
5. The new plasmid with the human insulin gene is put back into a bacterium. The bacterium has been genetically modified. The bacterium will now make human insulin.

Advantages and disadvantages of genetic engineering

✓ Making human insulin using GM bacteria is quicker and cheaper than producing it any other way. So more people with diabetes can be treated.

✗ A few diabetic people react badly to this insulin and need a different form.

✓ 'Golden rice' is a GM plant that contains genes from other plants that increases the production of beta-carotene. Beta-carotene makes vitamin A. It could prevent illness caused by lack of vitamin A in people who mainly eat rice.

✗ Golden rice seed costs more than normal rice seed, so the poorest people can't afford to grow it.

Worked example

A herbicide is a chemical that kills plants.

Crop plants can be genetically modified with a gene that protects them from being killed by a herbicide. Describe some advantages and disadvantages of growing GM crops.

Advantages: Only the weed plants will die, so the crop plants will have more room to grow. This should increase the crop yield.

Disadvantages: If the gene for herbicide resistance passes to weed plants, the weeds will no longer be killed by the herbicide. So the cost of making GM crop plants will be wasted. Also, overuse of herbicides can reduce biodiversity.

Now try this

target E-C

1. Describe what we mean by a genetically modified organism. **(2 marks)**
2. Describe how 'golden rice' plants could be produced. **(3 marks)**

target D-B

3. Some people are against genetic engineering for ethical reasons. Give two scientific arguments against producing genetically modified crops. **(2 marks)**

Mitosis

There are two types of cell division. Mitosis is the cell division that happens in body cells. A body cell is any cell except those that produce gametes (sex cells).

Three things to remember

- Mi-to-sis makes TWO cells
- MiTosis makes idenTical cells
- Diploid means Double (two sets of) chromosomes

The cell that divides is called the parent cell.

The parent cell divides to form two daughter cells.

The parent cell is a diploid cell. This means it has two sets of chromosomes.

Before the parent cell divides, each chromosome is copied exactly.

When the cell divides in two, each cell gets one copy of each chromosome.

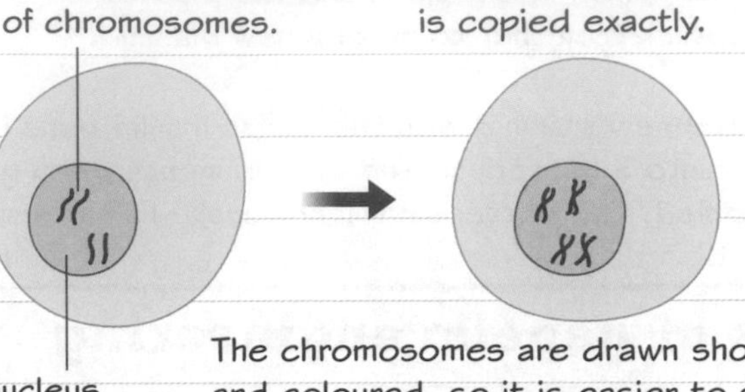

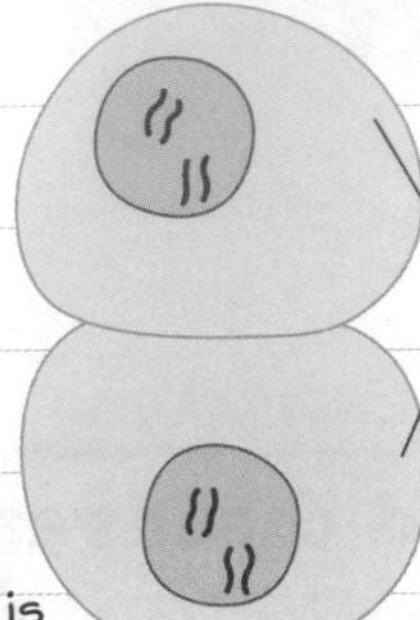

The daughter cells are genetically identical to each other and the parent cells. They are also diploid cells.

The chromosomes are drawn short here, and coloured, so it is easier to see what is happening. They don't really look like this.

Worked example

Name three situations where mitosis is used to produce new cells.

Mitosis produces new body cells for:

- growth
- repairing damaged parts of the body
- asexual reproduction.

Remember: asexual reproduction produces genetically identical offspring.

Asexual reproduction

Asexual reproduction is the production of new organisms without fertilisation, such as when:

- bacteria split in two to make more bacteria
- plants make new plantlets that split off from the parent plant to grow on their own.

Now try this

1. Explain what is meant by:
   (a) diploid
   (b) a daughter cell. **(2 marks)**
2. Explain why cells produced by mitosis are genetically identical diploid cells. **(3 marks)**
3. If several plants are produced by asexual reproduction from the same parent plant, will their flowers all be the same colour or a range of colours? Explain your answer. **(2 marks)**

Fertilisation and meiosis

Fertilisation takes place during sexual reproduction. Fertilisation is when a male gamete (sex cell) combines with a female gamete to produce a zygote.

Gametes are haploid cells. They have only one set of chromosomes. → The zygote has two sets of chromosomes. So it is a diploid cell.

Sperm cell (gamete) carries chromosomes from the father.

Egg cell (gamete) carries chromosomes from the mother.

Gametes fuse at fertilisation.

not to scale: the egg cell is about 20 times larger than the head of a sperm cell

A zygote is formed, which has one set of chromosomes from the mother and one set from the father.

Remember: sexual reproduction produces variation in the offspring. Asexual reproduction produces offspring that are genetically identical.

Meiosis

Meiosis is the second type of cell division. It happens when a diploid cell divides to produce haploid gametes (sex cells, such as sperm and egg cells).

Remember: haploid cells, produced by meiosis (me-1-osis), have 1 set of chromosomes.

The parent cell is a diploid cell. So it has two sets of chromosomes. → The parent cell divides in two and then in two again. Four daughter cells are produced.

one set of chromosomes

the other set of chromosomes

pair of chromosomes

Before the parent cell divides, each chromosome is copied.

Each daughter cell gets a copy of one chromosome from each pair.

Each daughter cell has only one set of chromosomes. So these are haploid cells. The daughter cells are not all identical – meiosis results in variation.

EXAM ALERT!

Be careful not to confuse mitosis with meiosis. In a recent exam question on this subject fewer than one in ten students got full marks.

Students have struggled with this topic in recent exams - **be prepared!** ResultsPlus

The cells produced by division are always called 'daughter cells' even if they will eventually turn into sperm cells.

Now try this

1. Outline the process of fertilisation, using the words **diploid**, **gamete**, **haploid** and **zygote** in your answer. **(4 marks)**
2. Give two differences between mitosis and meiosis. **(4 marks)**

3. Describe the importance of meiosis occurring before fertilisation. **(3 marks)**

Clones

Clones are organisms that have identical genes. Cloning is an example of asexual reproduction.

Cloning mammals

To clone a mammal (such as a sheep or cat), you start with a body cell from an adult mammal and an egg cell from the same species. The clone will have exactly the same genes as the animal from which the body cell was taken.

nucleus removed from egg cell
The diploid nucleus is removed from an adult body cell.
The diploid nucleus from the body cell is placed inside the empty egg cell. The cell is then stimulated with an electric pulse to start mitotic division.
The cell divides and grows as an embryo.
The embryo is placed in the uterus of a surrogate mother until it is ready to be born.
The surrogate mother is a different animal from the donors of the body cell and egg cell.

Advantages and disadvantages

- ✓ If the animal that is cloned has good features, all of its offspring will have the same good features.
- ✗ It is more difficult to clone a mammal than a plant. It may take many attempts before a healthy cloned mammal is born, and each attempt costs more money.
- ✗ Cloned mammals may suffer more health problems than usual, which may cause them to die early.
- ✗ Any genetic defect in the parent will be passed on to the offspring.

Worked example

A farmer has a prize bull. Other farmers buy sperm from her bull to fertilise their cows to produce high-quality calves. The farmer considers having the bull cloned. State one advantage and one disadvantage of cloning the prize bull.

Advantage: The farmer will have more bulls that produce sperm which other farmers want to buy, so she will get more money.

Disadvantage: If it takes too many attempts to clone the bull, the farmer may spend more money than she can gain from selling the sperm.

Now try this

1. Define the term 'clone'. **(1 mark)**
2. Genetic modification is used to produce GM goats that have human hormones in their milk. The GM goats are then cloned.
   (a) Explain the advantage of producing GM goats by cloning instead of sexual reproduction. **(3 marks)**
   (b) State one risk of producing goats by cloning. **(1 mark)**

Stem cells

Cells in an embryo are unspecialised. They divide to produce all the differentiated cells in the body, such as neurones and muscle cells. Once the cells have differentiated they cannot divide to produce other kinds of cell.

Stem cells are cells that can divide to produce many types of cell. There are two kinds of stem cell:

- Embryonic stem cells are taken from embryos that contain only a few cells.
- Adult stem cells are found in differentiated tissue, such as bone or skin – they divide to replace damaged cells.

Worked example

Describe how embryonic stem cells change as an animal matures.

Embryonic stem cells can differentiate into almost any type of cell. As the animal matures, most cells lose this ability. Only a few adult stem cells remain that can differentiate into a small range of different types of cell.

Research using stem cells

Some diseases are caused by faulty cells, such as cystic fibrosis. Scientists are researching how to use stem cells to produce replacement healthy cells for treating these diseases.

embryonic stem cells
- ✓ easy to extract from embryo
- ✓ produce any type of cell
- ✗ embryo destroyed when cells removed – some people think embryos have a right to life
- ✗ body recognises the cells as 'different' and will reject them without use of drugs

all stem cells
- ✓ replace faulty cell with healthy cell, so person is well again
- ! may produce cancer cells instead of healthy cells

adult stem cells
- ✓ no embryo destroyed so not an ethical issue
- ✓ if taken from the person to be treated, will not cause rejection by the body
- ✗ difficult to find and extract from tissue
- ✗ produce only a few types of cell

✓ – Advantage ✗ – Disadvantage ! – Risk

EXAM ALERT!

An exam question on a recent paper asked a question on this subject and around half of all students did not get all the marks available. If an exam question asks for two benefits or advantages of research, make sure you give two *different* examples.

Students have struggled with exam questions similar to this - **be prepared!** ResultsPlus

Now try this

1. State what is meant by a stem cell. **(1 mark)**
2. Explain one disadvantage of research into embryonic stem cells. **(2 marks)**

3. **(a)** State why embryonic stem cells could be more useful than adult stem cells to replace faulty cells. **(1 mark)**
   (b) Describe one practical advantage of using adult stem cells from the patient instead of embryonic stem cells to replace faulty cells. **(2 marks)**

Protein synthesis

Protein synthesis consists of two separate stages: transcription and translation. Transcription takes place in the nucleus. The strand of messenger RNA (mRNA) that is formed then moves out of the nucleus into the cytoplasm. Translation takes place in the cytoplasm when the mRNA strand attaches to a ribosome.

Transcription

1. A section of DNA is unwound and the two strands separate.
2. Free complementary bases pair with the open bases on one DNA strand. The free bases are joined together to make a strand of complementary mRNA.

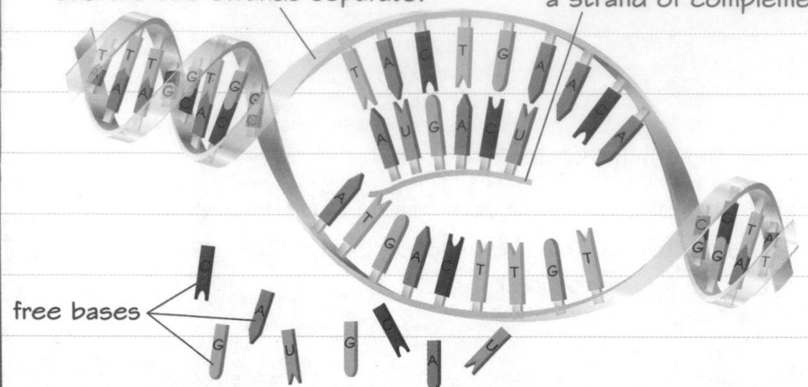

The base pairs that produce the strand of mRNA are the same as in DNA, except that T in DNA is replaced by U in RNA. So A and U pair, and G and C pair.

Translation

A polypeptide is one piece of a protein.

1. Ribosome moves along the mRNA in this direction reading one triplet of bases (codon) at a time.
2. tRNA molecules bring amino acids to the ribosome. The amino acid attached to each tRNA molecule depends on the order of bases in the tRNA.
3. Complementary bases of tRNA pair with the bases on the mRNA strand.
4. Amino acids that are close together are joined to make an amino acid chain (a polypeptide.)
5. tRNA free to collect another amino acid
6. Every protein is formed from a specific number of amino acids in a particular order. The order of the bases in the DNA defines the order in which the amino acids are joined together. So one section of DNA codes for one particular protein.

amino acids

C C G

C U A

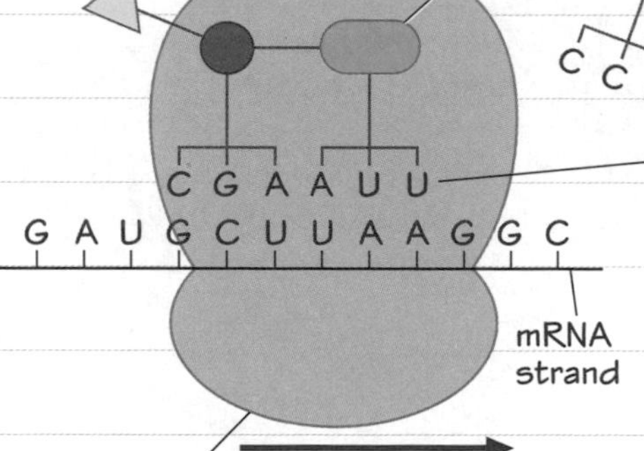

mRNA strand

To *transcribe* means to copy – the base order in DNA is copied to make the base order in mRNA. To *translate* is like changing to another language – translating from bases to amino acids.

Now try this

target D-C

1. Name the two stages in protein synthesis. **(2 marks)**

target D-B

2. Describe the role of the following in protein synthesis:
   (a) mRNA
   (b) tRNA. **(2 marks)**

target C-A

3. Explain the link between the order of the bases on the DNA and the order of the amino acids in a polypeptide. **(3 marks)**

Proteins and mutations

Each protein has its own specific number and sequence of amino acids. The sequence of amino acids gives each protein a particular 3D shape. The shape of a protein affects the way it works.

Shape	Examples	Functions
globular (blobby)	• enzymes • hormones • haemoglobin (in blood)	• control reactions due to their shape • transported in blood to target cells • carries oxygen in red blood cells
fibrous (long, strong fibres)	• keratin (in nails) • collagen (in tendons and ligaments)	• tough nails protect finger and toe ends • ligaments hold bones together, tendons attach muscles to bones

Mutations

A mutation can change one or more of the bases in the DNA base sequence. This may change the amino acid that is added to the chain during translation on the ribosome.

A change in amino acid may:

- have no effect if it doesn't change the shape of the protein or the way it works
- be beneficial if it helps the protein to work better than before
- be harmful if the protein changes shape and does not work so well. An example of a harmful mutation is the allele that causes sickle cell disease.

The allele that causes sickle cell disease has one base different than the normal allele for haemoglobin.

Worked example

Enzymes have a very specific shape that allows them to work effectively. Explain why a gene mutation can change the activity of an enzyme. Refer to the active site in your answer.

A gene mutation can change the base sequence in DNA. If this change produces a different amino acid sequence, this may change the shape of the active site. The shape of the active site controls how well an enzyme works. If the shape of the active site matches the shape of the substrate better, then the enzyme will work better. But if the shape of the active site is not as good a match, then the enzyme will not work so well.

Now try this

target D-C

1. Define the term **mutation**. **(1 mark)**

target D-B

2. Explain why different proteins have different shapes. **(2 marks)**

target C-A

3. Explain how a mutation in the gene that codes for keratin could result in brittle fingernails. **(3 marks)**

Enzymes

Enzymes are biological catalysts. Enzymes control reactions in the body.

This means they are found in living organisms.

A catalyst changes the rate of a reaction but remains unchanged at the end of the reaction.

Each enzyme is specific for its substrate, which means it only works with that substrate.

Some enzyme-controlled reactions occur inside cells to join substrate molecules together. For example:

- DNA replication

One enzyme catalyses the splitting apart (unzipping) of the two DNA strands.

A different enzyme catalyses the joining together of bases to make new strands.

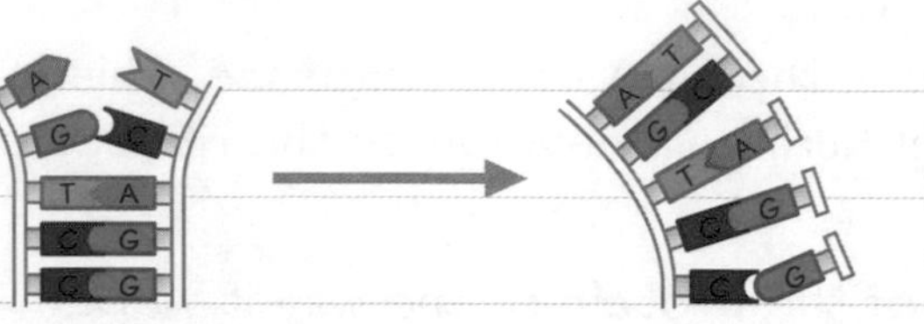

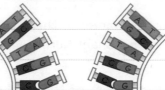

- Each reaction during transcription and translation in protein synthesis is controlled by a different enzyme.

Some enzyme-controlled reactions outside cells break substrate molecules down into smaller molecules. For example, digestive enzymes are secreted into the alimentary canal to break down the large molecules in our food into much smaller ones. The small food molecules can then be absorbed into the body.

Worked example

Label the diagram to describe the role of the active site in an enzyme-controlled reaction.

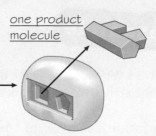

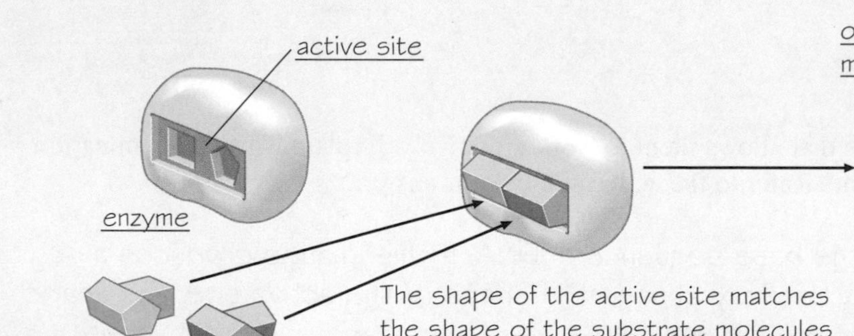

The shape of the active site matches the shape of the substrate molecules and holds them close together so bonds can form between them to make the product.

The product molecule doesn't fit the active site well so it is released from the enzyme.

Now try this

target E-C

1. Explain why enzymes can be described as biological catalysts. **(2 marks)**

target D-C

2. Give one example of an enzyme-controlled reaction inside that body that happens:
   (a) in cells
   (b) outside cells. **(2 marks)**

target C-A

3. Explain what is meant by the active site of an enzyme. **(2 marks)**

Enzyme action

'Lock and key'

The 'lock and key' hypothesis describes the way that the substrate fits like a 'key' into the active site (the 'lock').

This hypothesis explains why enzymes are specific (only work with one substrate), because only a substrate with the right shape can fit into the active site. The hypothesis also explains how enzyme activity changes in different conditions.

You can show the effect of changing a factor, such as temperature, on enzyme activity by using an enzyme that breaks down a substance that can be identified (e.g. starch) and measuring how quickly the substance is broken down at different values of the factor.

Temperature

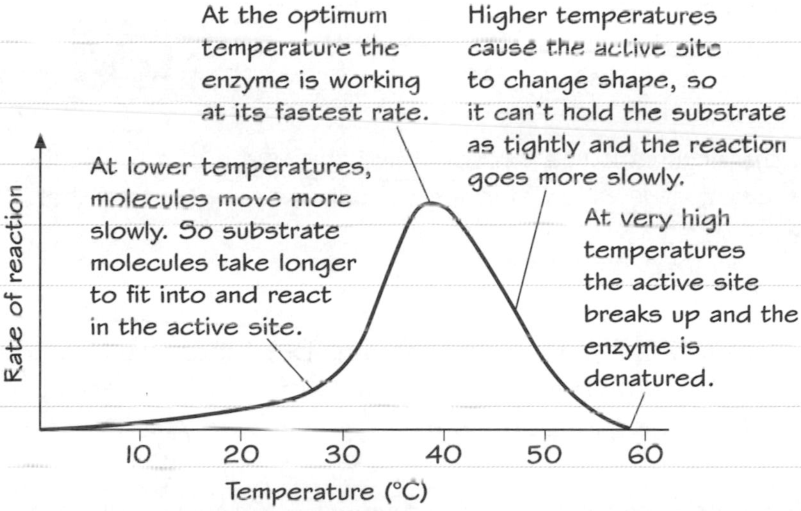

At the optimum temperature the enzyme is working at its fastest rate.

Higher temperatures cause the active site to change shape, so it can't hold the substrate as tightly and the reaction goes more slowly.

At lower temperatures, molecules move more slowly. So substrate molecules take longer to fit into and react in the active site.

At very high temperatures the active site breaks up and the enzyme is denatured.

Rate of reaction

10 20 30 40 50 60

Temperature (°C)

Substrate concentration

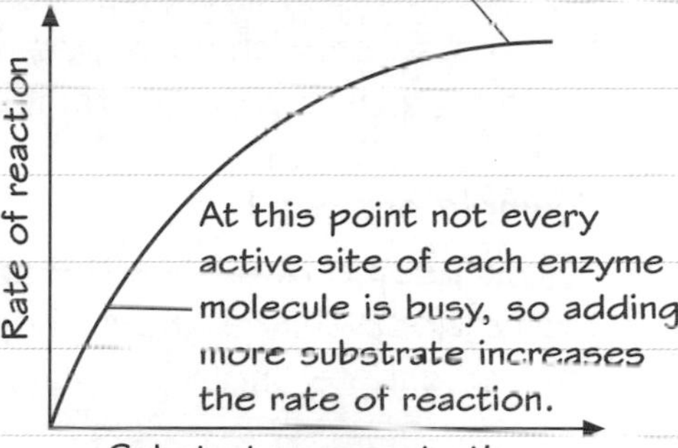

Adding more substrate at this point has little effect because the active site of every enzyme molecule is busy.
Rate of reaction
At this point not every active site of each enzyme molecule is busy, so adding more substrate increases the rate of reaction.
Substrate concentration

Worked example

Draw a graph of rate of reaction against pH for an enzyme with an optimum pH of 6. Label your graph to explain what you have drawn.

Changing the pH can change the shape of the enzyme's active site, and so change its ability to bond with the substrate.

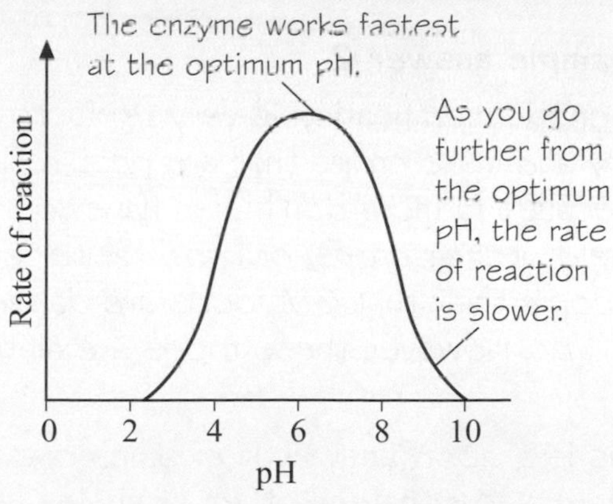

The enzyme works fastest at the optimum pH.
As you go further from the optimum pH, the rate of reaction is slower.
Rate of reaction
0 2 4 6 8 10
pH

Now try this

target D-B

1. Identify the optimum temperature for the enzyme in the temperature (top left) graph above. Explain your answer. **(2 marks)**

target C-B

2. Explain how the 'lock and key' hypothesis helps us understand why each enzyme only works with one substrate. **(3 marks)**

Biology extended writing 1

To answer an extended writing question successfully you need to:

- ✓ use your scientific knowledge to answer the question
- ✓ organise your answer so that it is logical and well ordered
- ✓ use full sentences in your writing and make sure that your spelling, punctuation and grammar are correct.

Worked example

Some people do not agree with the idea of producing food that has been genetically modified.
Compare the advantages and disadvantages of using genetic engineering to change organisms for human use. Use different examples of genetic engineering to help you answer the question. **(6 marks)**

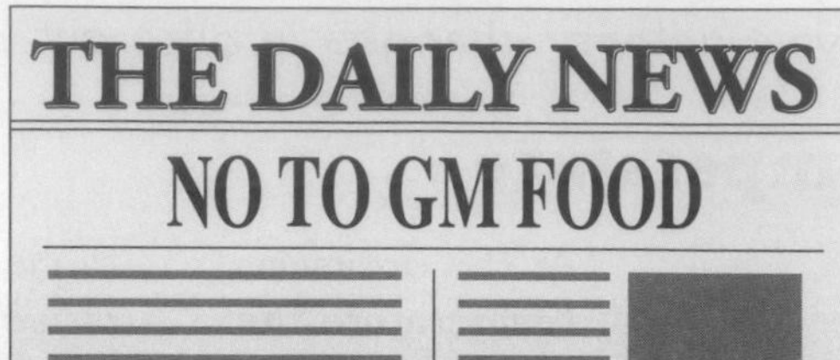

You need to present both sides of the argument in a question like this. Even if you find it hard to agree with an opposing viewpoint, it is good science to consider arguments that do not agree with your point of view.

Sample answer 1

Most people don't like the idea of GM foods because it's unnatural. But there must be some good things about them. You can make bigger food, so you can feed more people. But scientists shouldn't play God and change the food we eat.

This is a basic answer. It only really considers one side of the argument, giving common, non-scientific reasons for opposing GM foods. Although it says there may be some advantages, it doesn't give any details of what organisms can be engineered, or what this would produce.

Sample answer 2

Genetic modification is very useful to help make different foods. One example of this is to try and make crops that are pest-resistant by adding in the genes for this. This is good, because farmers don't then have to use chemicals to grow food. They can increase the yield of their crops, or grow tastier or better-looking versions of the same foods. Some people think that GM foods are dangerous because we don't know what the added genes will do. However, these foods are all tested so that we know they are safe.

This is a good answer. It explains one use of GM organisms in farming. A better answer would also consider other examples, such as the gene for making beta-carotene being inserted into rice; or using bacteria to produce insulin needed by diabetics. Only one disadvantage is discussed. Again, a better answer would consider other drawbacks such as the potential for loss of biodiversity or the potential dangers from the cross-breeding of 'normal' and GM products.

Now try this

1. In the human body, cell division can happen by mitosis or meiosis. Compare these two types of cell division, giving details of the two processes. **(6 marks)**

Biology extended writing 2

Worked example

Scientists at Seoul National University were the first to clone a dog, in April 2005. They called the clone Snuppy.

Explain the processes used to produce a clone like Snuppy. **(6 marks)**

Sample answer 1

A clone is an identical version of an animal, so this dog is identical to its parent. Scientists make a clone by fertilising an egg with DNA from the parent. This egg then grows to be a clone of the parent.

This is a basic answer. It correctly defines a clone, but does not give a good description of the process of cloning. In fact, this is a common misunderstanding – there is no fertilisation involved in the cloning process! To make the answer better you could set out all the steps in the process of cloning an animal.

Sample answer 2

Clones are identical copies of the parent. Cloning is a form of asexual reproduction. You need two things to start the process. The first is an egg cell. The nucleus is removed from this cell – this is called enucleation. You then take any body cell from the organism to be cloned, and take out the nucleus with all the DNA in it. This nucleus is put into the enucleated egg cell. This cell is then given an electric shock to start cell division going. The cell starts to divide and form an embryo and it can then be implanted into the womb of a surrogate mother.

This is an excellent answer. It uses some very good scientific terminology to describe the steps in cloning. The steps are presented in a clear, logical order.

Now try this

1. Describe how different forms of RNA take part in protein synthesis. **(6 marks)**

You might find it easier to make a list of what you know about the different forms of RNA and then link those ideas together to make short paragraphs.

Aerobic respiration

Respiration

All living organisms use respiration to release energy from organic molecules. The energy is used in the organism, for example, for growth and movement.

Aerobic respiration uses oxygen from the air to release energy from glucose. The products of aerobic respiration are carbon dioxide and water.

The reactions of aerobic respiration can be shown using a word equation:

glucose + oxygen → carbon dioxide + water (+ energy)

EXAM ALERT!

Make sure you know the difference between respiration (releasing energy in cells) and breathing (getting air into and out of the body). A question on this subject in a recent exam was only correctly answered by about four out of ten students.

Students have struggled with this topic in recent exams - **be prepared!** ResultsPlus

Diffusion

Many substances enter and leave the body by diffusion. These substances include gases such as oxygen and carbon dioxide, and small digested food molecules such as glucose.

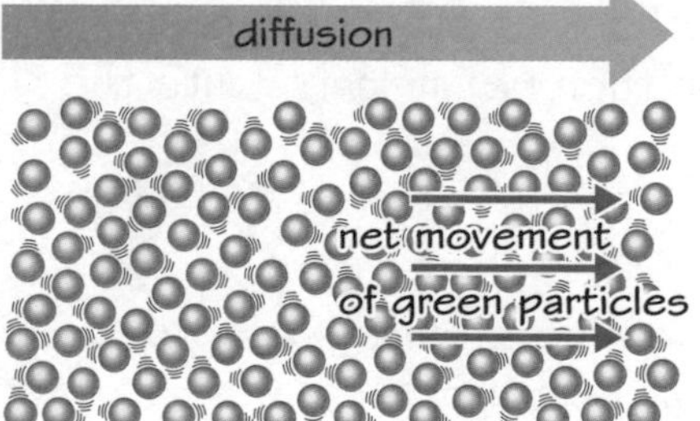

Diffusion is the net movement of particles from an area of higher concentration to an area of lower concentration. Particles are always moving and net movement is the sum of the movement of all particles.

Worked example

Complete the labels to show the role of the human circulatory system in supporting respiration.

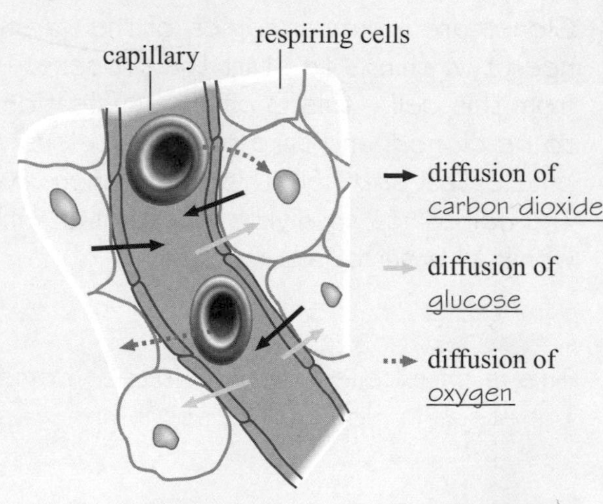

Now try this

target D-C
1. Write the word equation for aerobic respiration. **(1 mark)**

target D-B
2. Explain why respiration is important in living organisms. **(2 marks)**

target C-A
3. Describe the role of diffusion in supporting respiration in body cells. **(3 marks)**

Exercise

When you exercise, your heart rate and breathing rate increase. The harder you exercise, the more these rates increase.

Heart rate can be measured by taking your pulse at the wrist. It is usually measured as number of beats per minute. Breathing rate is measured by counting the number of breaths in one minute.

Heart rate and breathing rate can vary just by thinking about them. So take several measurements and average the results.

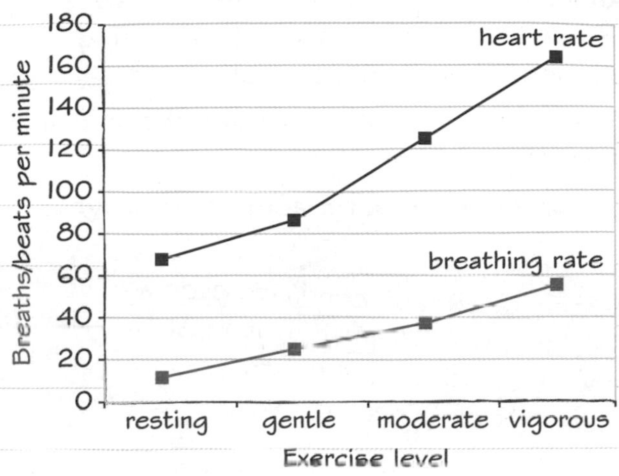

Why heart rate and breathing rate change

During exercise, muscle cells are respiring faster. This means that they need more oxygen and glucose, and release more carbon dioxide:

- A faster heart rate means that blood is pumped faster around the body. The blood takes oxygen and glucose to cells faster and removes carbon dioxide faster.
- A faster breathing rate means that oxygen can be taken into the body at a faster rate and carbon dioxide can be released faster.

Worked example

The table shows the stroke volume and heart rate for two people at rest.

	Stroke volume (cm^3)	Heart rate (beats per minute)
Trained athlete	90	55
Untrained person	60	70

Calculate the cardiac output for these two people.

Athlete: $90 \times 55 = 4950\, cm^3$ per minute

Untrained person: $60 \times 70 = 4200\, cm^3$ per minute

cardiac output = stroke volume × heart rate

Stroke volume is the volume of blood pumped by the heart in one beat.

EXAM ALERT!

Show all your working for a calculation.

Students have struggled with exam questions similar to this - **be prepared!** ResultsPlus

Now try this

1. Describe how heart rate varies with level of exercise. **(2 marks)**
2. Explain why breathing rate increases as level of exercise increases. **(2 marks)**
3. Suggest why a trained athlete may have a similar cardiac output to an untrained person but a lower resting heart rate. **(2 marks)**

Anaerobic respiration

Anaerobic respiration is the release of energy from glucose without using oxygen. This produces lactic acid.

Anaerobic respiration can be shown using a word equation:

glucose → lactic acid (+ energy)

Aerobic respiration continues even when the cell uses anaerobic respiration. It's just that anaerobic respiration releases the extra energy the cell needs but can't get from aerobic respiration.

EXAM ALERT!

A recent exam question on this subject was only correctly answered by about two-fifths of students. Remember: anaerobic respiration in muscle cells produces *only one substance* – lactic acid.

Students have struggled with this topic in recent exams - **be prepared!** ResultsPlus

Advantage

✓ Anaerobic respiration is useful for muscle cells because it can release energy to allow muscles to contract when the heart and lungs cannot deliver oxygen and glucose fast enough for aerobic respiration.

Disadvantages

✗ Anaerobic respiration releases much less energy from each molecule of glucose than aerobic respiration.

✗ Lactic acid is not removed from the body. It builds up in muscle and blood, and must be broken down after exercise.

EPOC

Anyone who exercises hard will find that their heart and breathing rate take a while to return to their normal resting rate when exercise has ended. This is known as excess post-exercise oxygen consumption (EPOC).

Breathing rate / Time graph: at rest; Start exercise; End exercise; End recovery.

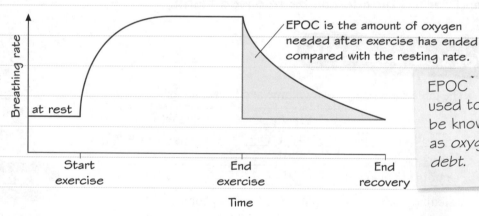

EPOC is the amount of oxygen needed after exercise has ended compared with the resting rate.

EPOC used to be known as oxygen debt.

Worked example

A student measured his resting pulse rate and breathing rate. Then he ran a 100 m sprint. After the race it took 5 minutes for his pulse and breathing rate to return to their resting rates. Explain why they didn't return to the resting rate as soon as the race finished.

Heart rate and breathing rate remain high after exercise to bring extra oxygen into the body. Some of this is needed to break down lactic acid produced from anaerobic respiration during the race.

Now try this

1. Explain what EPOC (excess post-exercise oxygen consumption) means.

(2 marks)

2. Explain why the EPOC of someone who has run fast for 10 minutes is greater than if they ran for 3 minutes.

(3 marks)

3. Explain why athletes who run for several hours must run at a speed that is mostly supported by aerobic respiration.

(3 marks)

Photosynthesis

Photosynthesis is the process that plants use to make glucose. During this chemical reaction, light energy is used to combine carbon dioxide and water.

The process of photosynthesis can be shown by this word equation:

carbon dioxide + water $\xrightarrow{\text{energy from light}}$ glucose + oxygen

The glucose made during photosynthesis is used by the plant for respiration and other processes. Photosynthesis takes place in the light. Respiration happens all the time.

Adaptations for photosynthesis

Photosynthesis takes place in cells that are mostly found in leaves. Many leaves have a large surface area to capture as much light as possible for photosynthesis.

Worked example

The diagram shows a section through a leaf. Explain how the labelled structures are adaptations that help the plant to photosynthesise.

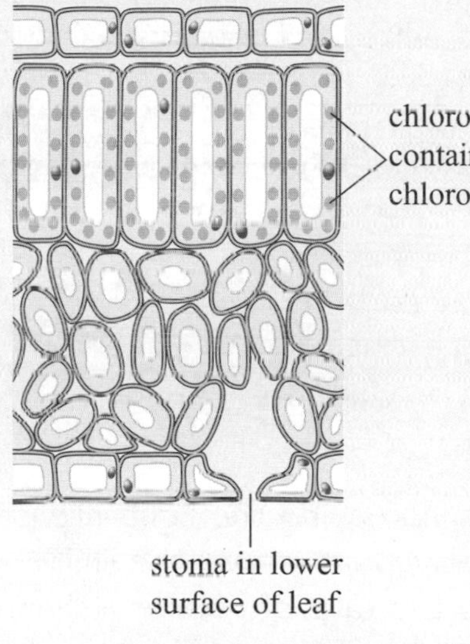

Chlorophyll in chloroplasts captures light energy needed for photosynthesis. The stomata in the leaf surface allow carbon dioxide needed for photosynthesis to diffuse into the leaf. They also allow oxygen and water vapour produced by photosynthesis to diffuse out of the leaf.

Remember: one **stoma**, two or more **stomata**.

You should mention diffusion when you are describing what stomata are for.

Now try this

target E-C 1. Identify the reactants and products of photosynthesis. **(2 marks)**

target D-B 2. State why photosynthesis usually only takes place in some leaf cells. **(1 mark)**

target C-A 3. Explain why most plants only open their stomata during the day. **(3 marks)**

Limiting factors

Low temperature, dim light and low carbon dioxide concentration all limit the rate of photosynthesis. They are all limiting factors for photosynthesis.

We can measure the effect of these factors on the rate of photosynthesis by measuring the rate at which oxygen is given off by a piece of pondweed.

You can use this apparatus to investigate the effect of:

- temperature using warm and cold water baths
- light intensity using bright and dim lights
- carbon dioxide concentration by adding different amounts of sodium hydrogen carbonate to the water.

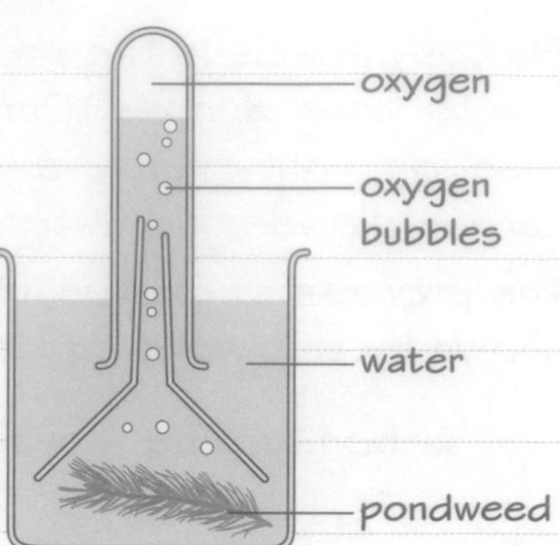

Worked example

The graph shows the rate of photosynthesis of a plant at different light intensities. Explain the shape of the graph at points A and B.

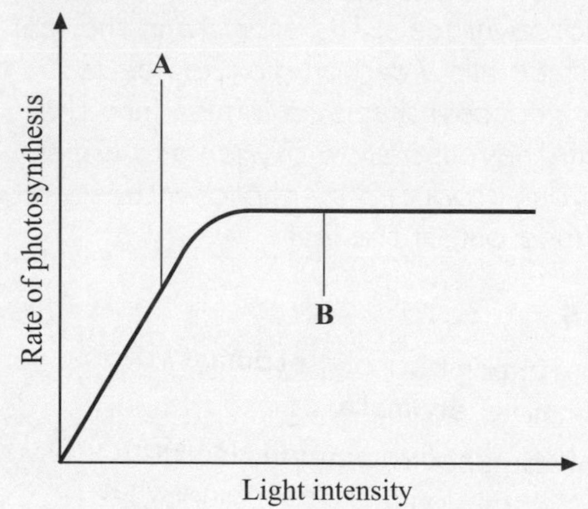

EXAM ALERT!

In a recent question that asked students to 'explain' how a factor affects plant growth, more than half lost marks because they failed to explain the effect of the factor on photosynthesis and just described how the plant growth changed.

Students have struggled with exam questions similar to this - **be prepared!** ResultsPlus

Increasing only carbon dioxide concentration or temperature, while keeping other factors constant, will produce a similar graph to this.

At A, light intensity is limiting photosynthesis, because increasing the light intensity increases the rate of photosynthesis. At B, light intensity is no longer the limiting factor because increasing the light intensity does not change the rate of photosynthesis. Another factor, such as carbon dioxide concentration or temperature, is limiting the rate of photosynthesis.

Now try this

target E-C

1. Define the term 'limiting factor'. **(2 marks)**

target D-B

2. Explain why measuring the oxygen given off by a plant is a way of measuring the rate of photosynthesis. **(2 marks)**

target C-A

3. Sketch and label a graph to explain the effect of temperature on the rate of photosynthesis from 0°C to 30°C. **(3 marks)**

Water transport

Water and substances dissolved in the water are transported around plants in veins. Veins contain tissues called xylem and phloem.

Glucose, produced by photosynthesis in leaves, is converted to sucrose, which is transported in phloem to the rest of the plant.

Water and dissolved minerals from the roots travel in xylem to the rest of the plant.

Water and dissolved minerals enter plants through their roots from the soil water.

Transpiration

Transpiration is the evaporation of water from inside leaves out into the air. It causes water to move up the plant from the roots.

evaporation of water vapour from leaves (transpiration)

↓

draws water out of the leaf cells and xylem

↓

draws water up the xylem from the roots

↓

causes water to enter the roots by osmosis

Worked example

Explain the meaning of the phrase *active transport*. Use minerals entering roots as your example.

Mineral salts cannot enter the root cells from soil water by diffusion because there is a higher concentration of mineral salts in the cells than in the soil. So the root cells have to use energy to transport mineral salts into the cells against their concentration gradient.

Types of transport

Diffusion
- substances move down their concentration gradient
- no energy needed

Active transport
- substances move up (or against) their concentration gradient
- needs energy so substances can move

Osmosis = diffusion of water molecules (next page)

Now try this

1. Describe how and where water and mineral salts enter a plant. **(2 marks)**
2. Describe the role of xylem and phloem in the transport of substances around a plant. **(2 marks)**

target C-A

3. Explain how transpiration causes water to move through a plant from soil to air. **(4 marks)**

Osmosis

Osmosis is the net movement of water molecules from a region of their higher concentration to a region of their lower concentration through a partially permeable membrane.

A partially permeable membrane is a membrane that lets some molecules through but not others. Cell membranes are partially permeable membranes because small molecules pass through but not bigger molecules.

Osmosis is a special case of diffusion – it is the diffusion of water.

Osmosis can be investigated using potato strips or red onion cells in solutions of different concentrations.

How the molecules are moving

A concentrated sugar solution contains many sugar molecules and some water molecules.

A dilute sugar solution contains many water molecules and some sugar molecules.

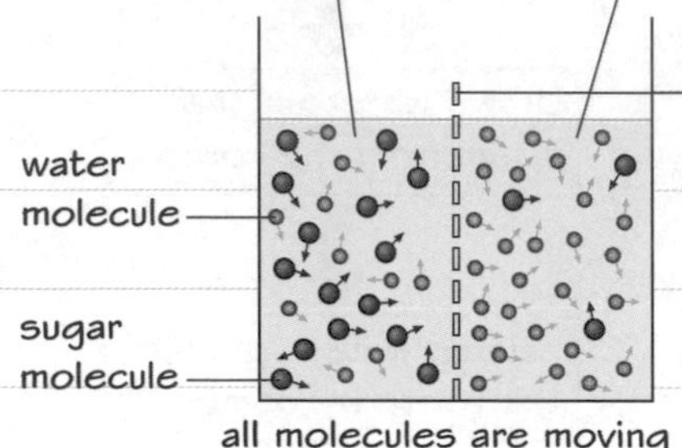

all molecules are moving

After a few minutes

Water molecules are small enough to pass through the membrane, while sugar molecules are too large.

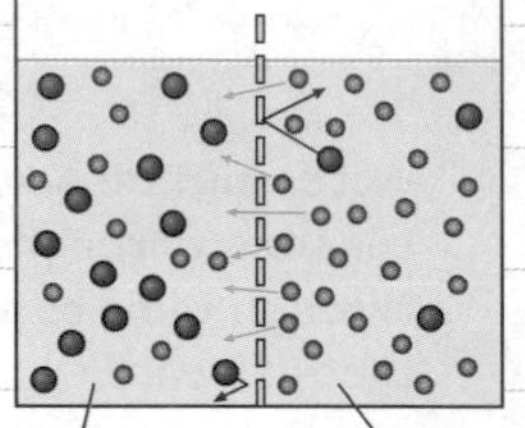

few water molecules means only a few cross through the membrane

lots of water molecules means lots cross through the membrane

After a while

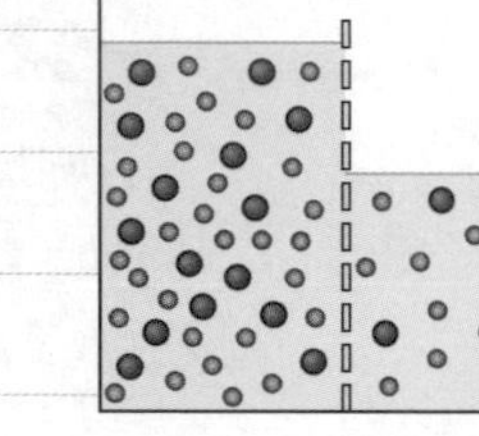

Worked example

The diagram shows two root hairs on the outside of a root. Describe how these cells are adapted to take up water by osmosis.

Root hair cells have long extensions that stretch out into the soil. This gives them a large surface area where osmosis can take place, which means that more water molecules can cross the cell membrane into the cell at the same time.

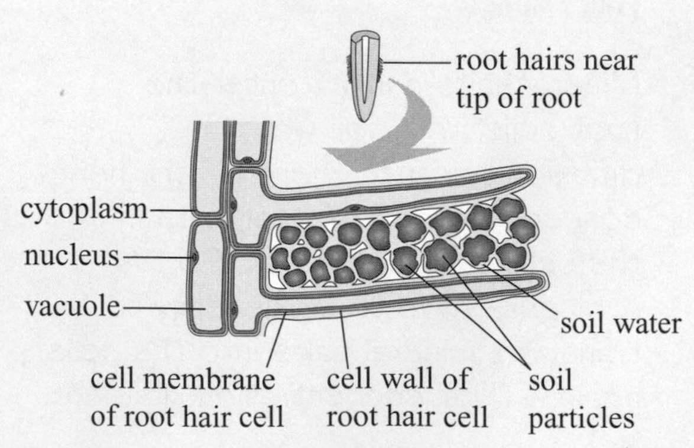

Now try this

1. Write a definition for **osmosis**. **(4 marks)**

2. Look at the diagram at the top of the page. Explain why the level of solution rises on one side of the membrane and falls on the other after a while. **(3 marks)**

3. Some potato strips were weighed and then placed in concentrated sucrose solution for 2 hours. Suggest how the weight of the potatoes is most likely to change after that time. **(3 marks)**

Organisms and the environment

When you carry out fieldwork to investigate the relationship between organisms and their environment, you need suitable equipment and sampling techniques.

Collecting animals

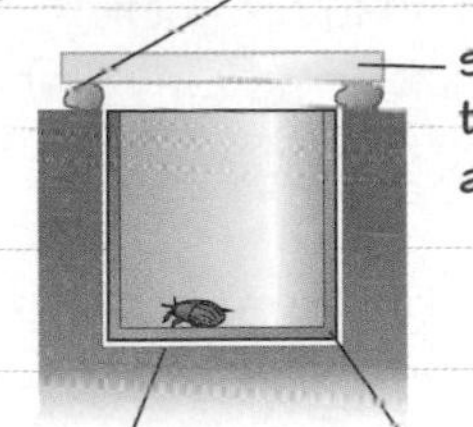

pitfall trap
for collecting small ground-living animals

sweep through grass and bushes or through water in pond or stream

sweep net or pond net
for collecting larger animals

inlet tube
mouth piece
a suck on the mouthpiece draws the animal in through the inlet tube
net to stop the animals getting into your mouth

pooter
for collecting animals smaller than the inlet tube

Quadrats

A quadrat is a square frame of a particular size (e.g. 50 cm²). A quadrat can be used to sample plants or animals that don't move much. To get good results:

- each quadrat should be placed randomly in the sample area
- count the number of organisms, or estimate the area covered by the organism, in each quadrat
- repeat the measurement in several quadrats
- calculate an average measurement for one quadrat.

Quadrats can also be used to sample organisms along a straight line – to see how the distribution of organisms changes along the line.

Environmental factors

Worked example

In a fieldwork investigation, describe how you would measure these environmental factors:
(a) temperature **(b)** light intensity **(c)** pH of soil.

(a) Temperature probe connected to a datalogger, or a thermometer.
(b) Light probe connected to a datalogger, or a light meter.
(c) Mix a sample of soil with distilled water, then measure the pH of the solution with Universal indicator or a pH probe.

Now try this

target D-C

1. Explain why it is good practice to sample using several quadrats in an area, and then average the results from each quadrat. **(2 marks)**

target D-B

2. Describe how you would use a quadrat to measure the population size of a species of snail in grassland. **(2 marks)**

target C-A

3. In a survey using 25 cm² quadrats, the following numbers of daisies were found in each quadrat: 3, 1, 2, 0, 5. Calculate the estimated population size of daisies in the meadow, which was 20 m × 15 m. **(4 marks)**

Biology extended writing 3

Worked example

Two students are looking at the distribution of beetles in the field behind their school.

They have a hypothesis that beetles will prefer to live in long grass rather than short grass.

Describe how they could carry out an investigation to collect and count beetles in the field to test their hypothesis. **(6 marks)**

Sample answer 1

They would look at some parts of the field. To do this, it would be best if they used a quadrat, so that they looked at the same area each time. They could collect the beetles that they found in each area, probably using a net. A tally chart would help them count the number of beetles in each quadrat. They should do the test more than once in each part of the field, so they could find an average number of beetles. They have to be careful to put the beetles back afterwards and not be cruel to them.

This is a good answer. There are several correct ideas, although it would be better if it went into a little more depth for each idea. For example, which parts of the field should be sampled? It's good that the answer mentions repeats, but remember that repeating an experiment allows you to remove anomalies, as well as finding an average. Some ideas of about controlling variables would improve the answer. The comment on ethical treatment of the organism is very good.

Sample answer 2

The students have to take samples from parts of the field with long and short grass. They should try and take more than one sample from each part, so that they can check their results. A quadrat would help them sample the same sized area each time – so it's a fair test. If they can, they should collect the data at the same time of day, and in the same conditions, so that whether it's a sunny day or a wet day doesn't affect their results. To collect the beetles, they could use pooters or a pit trap. These are both good because they do not harm the beetles. Once they have collected their results, they could plot a bar chart to show the numbers of beetles in each habitat.

This is an excellent answer. It gives some of the points that were missing from Answer 1. The key difference is stating where the sampling should take place. This is crucial because it will allow the students to relate their results back to the hypothesis. There are some good examples of controlling variables, and two safe collecting methods are considered. It is also good that the answer says how the data should be presented in order to test the hypothesis.

Now try this

1. Explain how plants take water out of the soil and transport it to their leaves. **(6 marks)**

Fossils and evolution

Fossils are the preserved traces or remains of organisms that lived thousands or millions of years ago. The fossil record in rocks gives us evidence of what the living organisms were like and how they changed over time. Change over time is called evolution.

Gaps in the fossil record

The fossil record is not complete. There are several reasons for this.

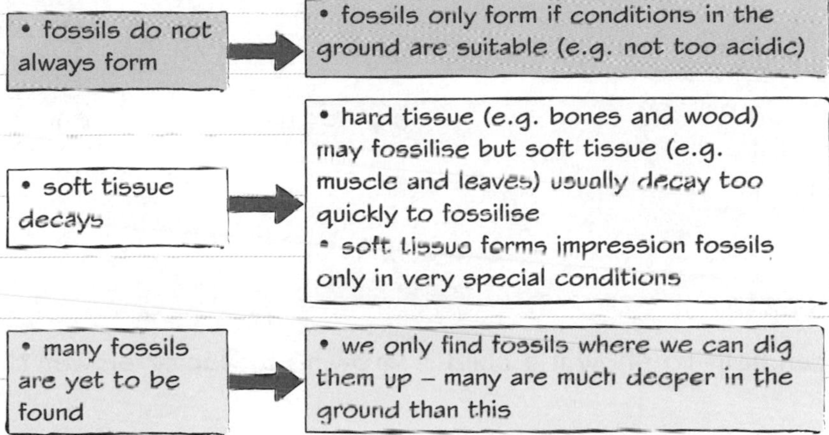

• fossils do not always form
• fossils only form if conditions in the ground are suitable (e.g. not too acidic)
• soft tissue decays
• hard tissue (e.g. bones and wood) may fossilise but soft tissue (e.g. muscle and leaves) usually decay too quickly to fossilise
• soft tissue forms impression fossils only in very special conditions
• many fossils are yet to be found
• we only find fossils where we can dig them up – many are much deeper in the ground than this

Worked example

The diagram shows the bones of the forelimbs of two living vertebrates. Explain how these pentadactyl limbs are evidence for evolution.

The limbs show the same basic pentadactyl structure. They have long bones supporting carpal bones and phalanges. This suggests that the human and the whale evolved from a shared ancestor. The differences between the limbs can be explained by evolution. The limbs have developed in different ways because they have adapted to different uses.

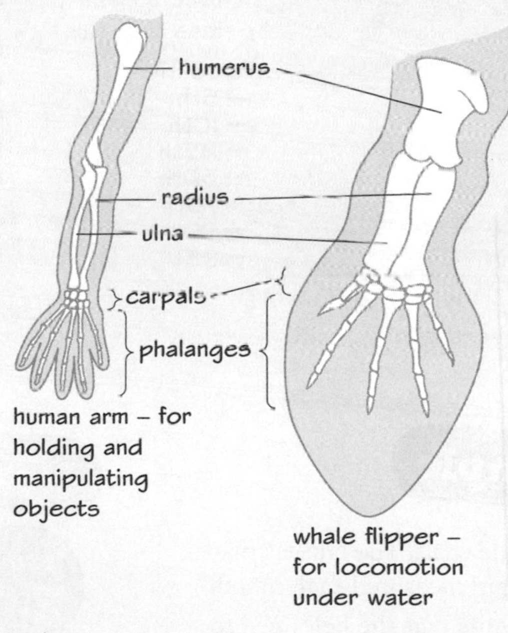

Many living vertebrates have the same basic pentadactyl (five-fingered) structure to their limbs.

Now try this

1. Explain why the fossil record is not complete. **(3 marks)**

2. Birds are the only living things with feathers. Scientists disagreed about which group birds evolved from. In 1996 the first feathered dinosaur fossil was found. Suggest how this fossil gave evidence for bird evolution. **(2 marks)**

Growth

When organisms grow they get bigger. Growth can be measured in different ways.

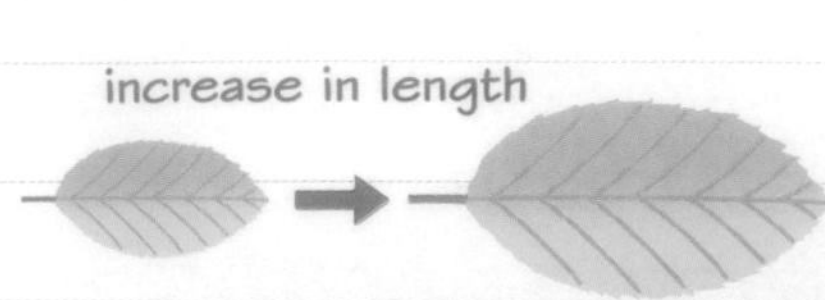

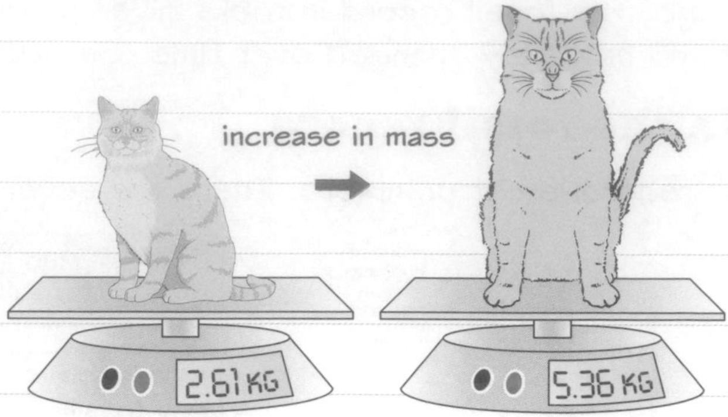

Growth is a *permanent* increase in size. For example, a balloon that is blown up a little more has not 'grown' in size.

Percentile charts

Percentile charts can help to show if a child is growing faster or slower than is normal for their age.

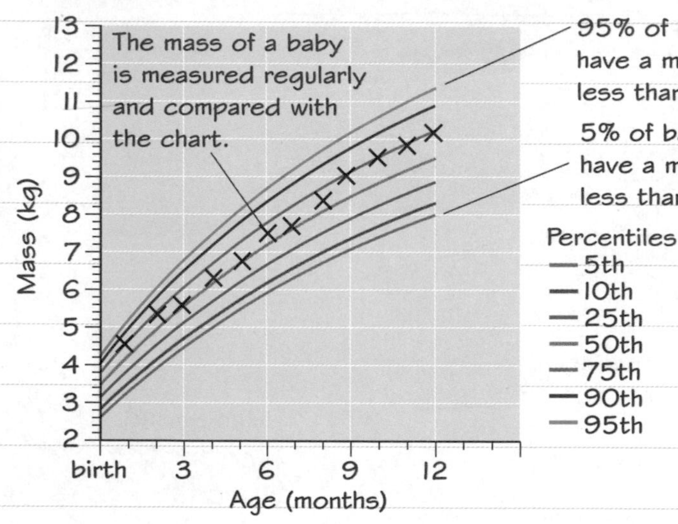

95% of babies have a mass less than this

5% of babies have a mass less than this

Percentiles
— 5th
— 10th
— 25th
— 50th
— 75th
— 90th
— 95th

Babies with a mass above the 95th line or below the 5th line may not be growing properly.

A baby whose mass decreases by two or more percentiles over their first year may not be growing normally.

The chart is for baby girls. The chart for baby boys is similar, but they grow at a slightly different rate to girls.

Worked example

Look at the percentile chart. The crosses mark the mass of a baby girl measured each month.

(a) State the percentile that she belonged to in her first month.

(a) 75th percentile.

(b) Do you think there was concern about this baby's mass increase in her first year? Explain your answer.

(b) No, because her mass varied only between the 75th and 50th percentiles, and this amount of variation is normal.

Now try this

1. Describe one way in which you could measure the growth of a plant. **(2 marks)**

2. Describe what a percentile chart is used for. **(2 marks)**

3. Explain why an increase in the size of a balloon is not an example of growth but increase in size of a child is. **(2 marks)**

Growth of plants and animals

Growth and development in plants

Plant growth includes cell division just behind the shoot tip, and cell elongation (the cells get longer) further away from the tip of the shoot. Cells at the tip of the shoot are all the same. They can develop into any kind of cell in a plant.

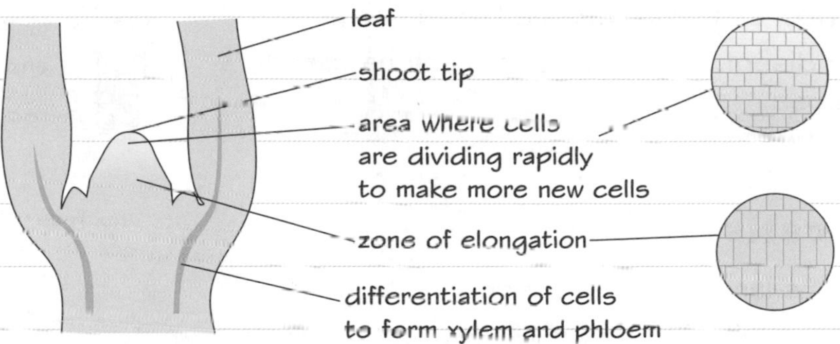

Growth and development in animals

Animals grow in a different way from plants. As the animal gets older, most of the cells in their body differentiate into specialised cells, such as muscle, bone or nerve cells.

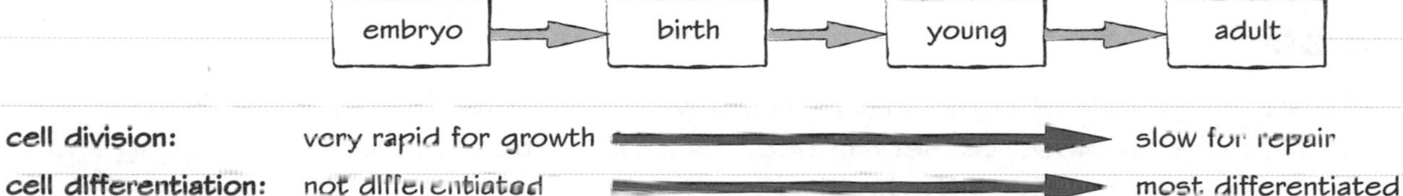

cell division: very rapid for growth → slow for repair

cell differentiation: not differentiated → most differentiated

All the cells in an embryo are stem cells. This means they can produce almost any kind of cell. Only a few cells in a child or adult's body are undifferentiated stem cells. These stem cells help to repair damaged tissue.

Worked example

Compare the growth and differentiation of cells in plants and animals.

Plants produce new cells at their shoot and root tips and continue growing all their lives. Differentiation of plant cells happens behind the shoot or root tips.

Growth by cell division in animals is rapid in the embryo but gets slower as the animal gets older. Cells differentiate as the animal gets older until only a few stem cells are left.

Now try this

1. State one similarity between plant cells in the tip of the shoot and animal stem cells. **(1 mark)**
2. Describe how cell division, elongation and differentiation contribute to the growth and development of a plant. **(3 marks)**

target D-A

3. Describe how cell division and differentiation contribute to the growth and development of an animal. **(2 marks)**

Blood

Blood is made of four main parts: plasma, red blood cells, white blood cells and platelets. Each part of blood has a particular function (job).

Blood plasma

Plasma is the liquid part of blood:

- It carries the blood cells through the blood vessels.
- It contains many dissolved substances, such as carbon dioxide and glucose.

plasma (55%)

white blood cells and platelets (<1%)

red blood cells (45%)

White blood cells

White blood cells are larger than red blood cells, and they have a nucleus. All types of white blood cells are part of the immune system, which attacks pathogens in the body.

nucleus

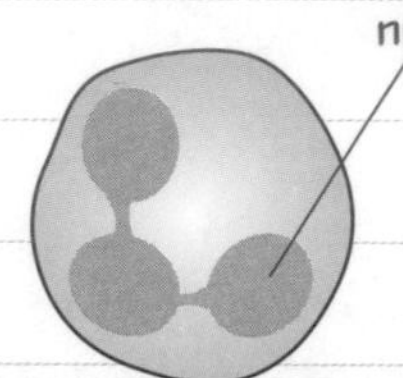

Some white blood cells surround and destroy pathogens.

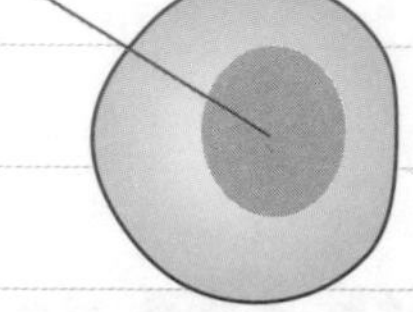

Some white blood cells produce antibodies that destroy pathogens.

Platelets

Platelets are fragments of larger cells. They have no nucleus. Their function is to cause blood to clot when a blood vessel has been damaged. The clot blocks the wound and prevents pathogens getting into the blood.

Worked example

Explain how the structure of a red blood cell is related to its function.

Biconcave means the cell is dimpled on both sides so that it is thinner in the middle than at the edges.

Red blood cells contain haemoglobin which carries oxygen. The biconcave shape of a red blood cell means it has a large surface area. This makes it easier for oxygen to diffuse into and out of the cell. The cell has no nucleus. This means the cell has room for more haemoglobin to carry more oxygen.

Now try this

target E-C

1. Respiring cells need oxygen and glucose. State which parts of the blood carry each of these substances. **(2 marks)**

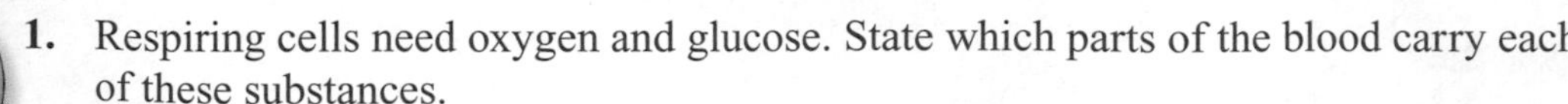

target D-B

2. Describe two ways in which white blood cells help to protect the body against disease. **(2 marks)**

target C-A

3. Explain how platelets help to protect the body against infection. **(3 marks)**

The heart

Cells are grouped into tissues and tissues are grouped into organs. The heart is an organ that contains tissues such as heart muscle and tendons. Heart muscle is formed from heart muscle cells.

Structure and function of the heart

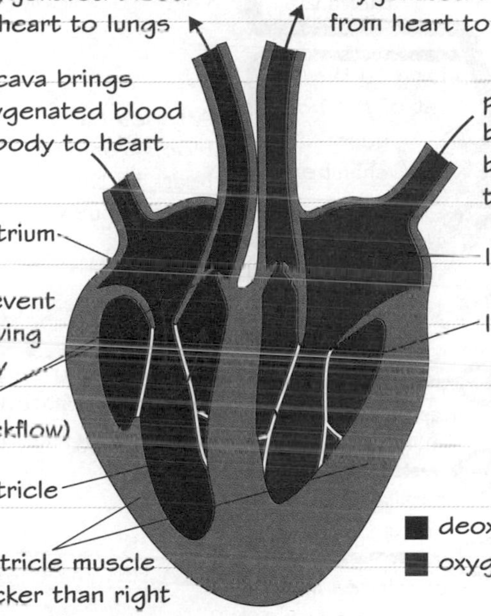

pulmonary artery carries deoxygenated blood from heart to lungs

aorta carries oxygenated blood from heart to body

vena cava brings deoxygenated blood from body to heart

pulmonary vein brings oxygenated blood from lungs to heart

right atrium

left atrium

valves prevent blood flowing wrong way through heart (backflow)

left ventricle

right ventricle

left ventricle muscle wall thicker than right ventricle as it pushes blood all round the body

■ deoxygenated blood
■ oxygenated blood

The sides of the heart are labelled left and right as if you were looking at the person. So the left side of the heart is on the right side of the diagram.

Remember - arteries take blood away from the heart, veins bring it back in to the heart.

Worked example

Starting at the vena cava, identify the chambers and blood vessels that deoxygenated blood flows through to the lungs.

Deoxygenated blood from the body enters the right side of the heart through the vena cava. It is pumped through the right atrium and then through the right ventricle, and leaves the heart through the pulmonary artery to reach the lungs.

Oxygenated blood contains a higher concentration of oxygen than deoxygenated blood.

Now try this

target D-C

1. Using the heart as an example, identify one tissue and one cell type. **(2 marks)**

target D-B

2. (a) Describe the role of valves in the heart. **(1 mark)**
   (b) Explain why the left ventricle wall contains more muscle than the right ventricle wall. **(2 marks)**

target B-A

3. Suggest an advantage of having a heart that keeps deoxygenated and oxygenated blood separate. **(2 marks)**

The circulatory system

An organ system is a group of organs that work together to carry out a particular function in the body. The circulatory system is an organ system that consists of the heart, the blood vessels and the blood. Its function is to transport materials around the body.

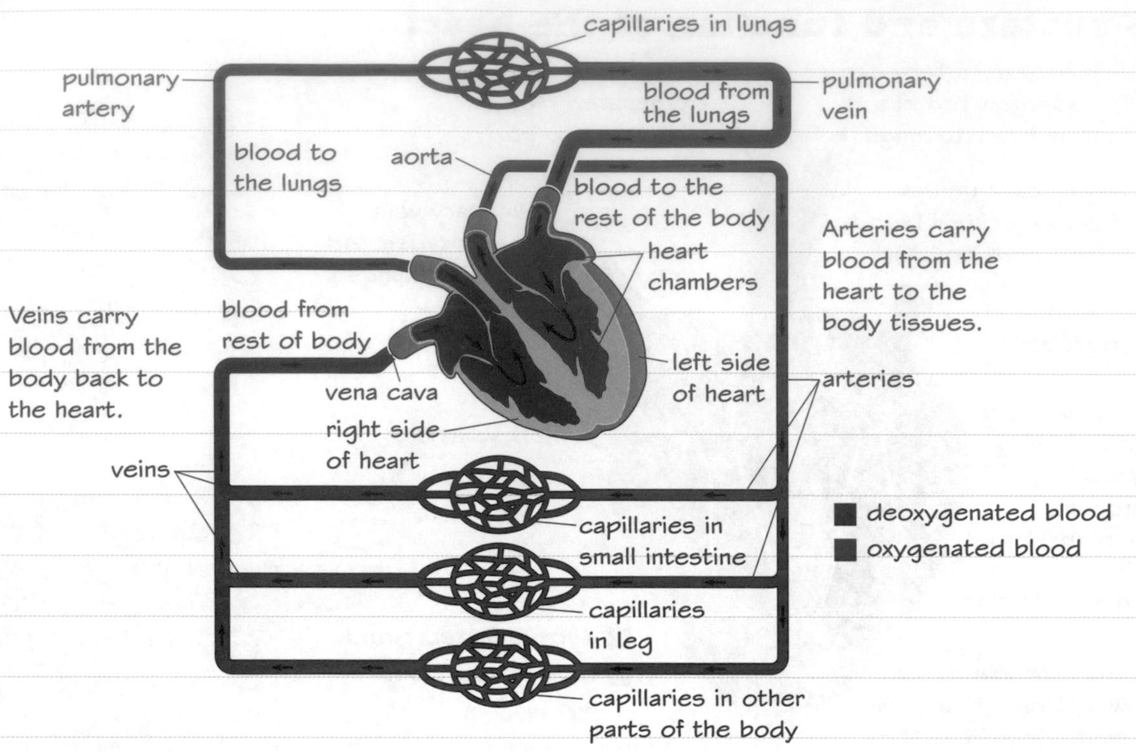

Worked example

Describe the role of arteries, veins and capillaries in the human circulatory system.

Arteries carry blood away from the heart – all arteries except the pulmonary arteries carry oxygenated blood to the body. Veins carry blood towards the heart – all veins except the pulmonary veins carry deoxygenated blood from the body back towards the heart. Capillaries exchange materials, such as oxygen, glucose and carbon dioxide, with body tissues.

Remember: a key function of blood is to deliver oxygen and glucose to cells for respiration and to remove the carbon dioxide produced by respiration from cells.

Now try this

1. Describe the role of the circulatory system in the body. **(1 mark)**
2. State the difference between the role of arteries and veins in the circulatory system. **(2 marks)**

The digestive system

There are many organs in the digestive system. These organs work together to digest food and allow food molecules to be absorbed into the body.

mouth
- food enters the body
- chewing breaks up food and mixes it with enzymes to start digestion

oesophagus
- carries food from mouth to stomach

liver
- produces bile, which helps in digestion of fats in small intestine
- converts food molecules absorbed from small intestine into other molecules

gall bladder
- where bile from liver is stored until needed in the small intestine

small intestine
- digestion of food by enzymes completed
- food molecules are absorbed into the blood
- water absorbed from digested food

food moving down oesophagus
- food moves through the oesophagus by peristalsis – peristalsis is the contraction of some muscles in the wall of the oesophagus, and relaxation of others, which pushes the food along from the oesophagus to the anus

stomach
- acid and enzymes added
- stomach contents mixed by churning of muscular wall

pancreas
- produces enzymes that are released into the small intestine

large intestine
- some water absorbed
- undigested food forms faeces that pass out of body through anus

anus

Worked example

Explain the role of enzymes and bile in the digestion of food molecules.

Different enzymes digest different food molecules so that they can be absorbed:

- carbohydrases digest carbohydrates, e.g. amylase digests starch to simple sugars
- proteases digest proteins to amino acids
- lipases digest fats to fatty acid and glycerol.

Bile is added in the small intestine to neutralise stomach acid. This means that the enzymes in the small intestine can work at their optimum pH. Bile also emulsifies fats, so that there is a greater surface area for lipases to work on.

To emulsify means to break up large drops of fats into much smaller droplets that remain mixed in a watery liquid.

Now try this

1. The two main functions of the digestive system are digestion and absorption. Identify which organs carry out these functions. **(3 marks)**
2. Describe two roles of the liver in the digestive system. **(2 marks)**
3. Describe why bile is important in the digestive system. **(4 marks)**

Villi

The surface of the small intestine is adapted for efficient absorption of the soluble products of digestion.

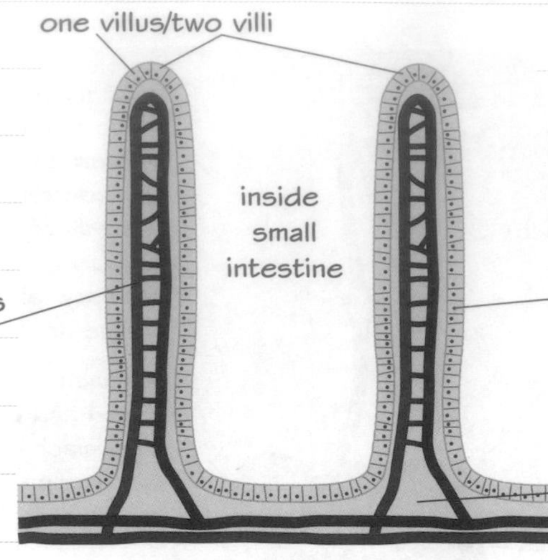

Visking tubing as a model

Visking tubing can be used in experiments to investigate digestion, including the effect of different concentrations of digestive enzymes on the digestion of large food molecules.

Worked example

Use the diagram to help you explain why Visking tubing can be used as a model for the alimentary canal.

In the diagram the inside of the Visking tubing is like the inside of the alimentary canal. The Visking tubing is like the wall of the small intestine because small molecules can diffuse through it but not large ones.

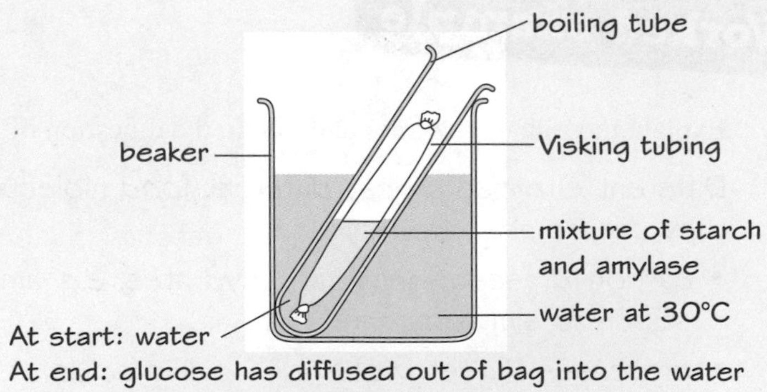

Now try this

1. (a) State three ways in which the structure of small intestine wall maximises the rate of absorption of small food molecules. **(3 marks)**

   (b) Explain why each of the answers given in part (a) maximises the rate of absorption. **(3 marks)**

If you are writing about the absorption of food molecules in the small intestine, remember to consider the role of diffusion and the factors that affect it.

Probiotics and prebiotics

Functional foods are foods that claim to make you healthier when you eat them.

Probiotics

Probiotic foods contain 'friendly' bacteria that are usually either *Bifidobacteria* or *Lactobacillus*.

Manufacturer claim:
Probiotic foods make you healthier by improving your digestive system and immune system.

Scientific studies show:
There is little or no evidence to support these claims.

Prebiotics

Prebiotic foods contain high levels of oligosaccharides.

Manufacturer claim:
Prebiotic oligosaccharides encourage beneficial bacteria to grow in your alimentary canal, which can protect you from problems such as diarrhoea.

Scientific studies show:
There is increasing evidence that prebiotics improve health by reducing the risk of diarrhoea and other problems.

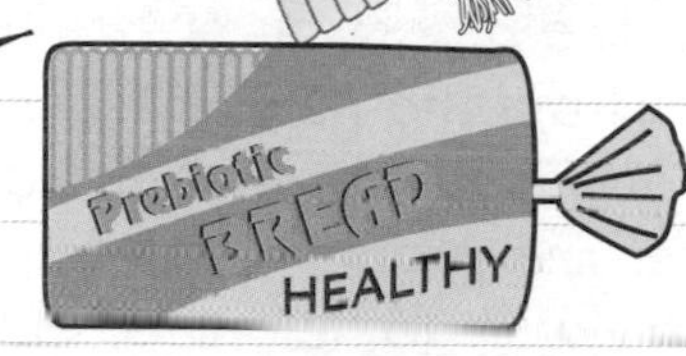

Worked example

Plant stanol esters are used as functional foods in yogurt, drinks and spreads. Explain what plant stanol esters do in the body to make you healthier.

Plant stanol esters reduce the amount of cholesterol absorbed from digested food. Reducing the amount of blood cholesterol may reduce your chance of having heart disease.

There is evidence that plant stanol esters reduce the amount of cholesterol in the blood. Scientists are still testing whether they also reduce the risk of heart disease.

Now try this

target D-C

1. Describe what is meant by a functional food. **(1 mark)**
2. Describe how prebiotic foods are supposed to make you healthier. **(1 mark)**

target B-A

3. Explain why scientific evidence is important when evaluating functional foods. **(2 marks)**

Biology extended writing 4

Worked example

Explain how different parts of the digestive system are involved in the digestion of fats. **(6 marks)**

Sample answer 1

The acid in the stomach churns up our food and starts breaking up fats. In the intestine, there are lots of enzymes. One of these enzymes is called lipase. It splits up fats into smaller parts, called fatty acids. These are then absorbed further down the intestine.

This is a basic answer. Some of the information is not correct – the stomach is where protein digestion starts, not digestion of fats. In other places, the answer is not detailed enough – there needs to be detail about the role of the 'intestine'. However, the information about lipase being broken into fatty acids is correct. A better answer would correct these areas, but would also give information about other parts of the digestive system that are involved.

Sample answer 2

Most of the digestion of fats happens in the small intestine. This is where all the enzymes from the pancreas start working. The key enzyme for digesting fat is called lipase. Once fats are broken down, then they can be absorbed further down the small intestine. The problem with fats is that they don't mix with the water in the digestive tract very well. To help this, the gall bladder makes a substance called bile. This helps the fats to mix with water.

This is a good answer, but it does contain one common error. Bile is made in the liver, and only stored in the gall bladder. There are some key ideas missing from this answer – such as the fact that fats are broken down into fatty acids and glycerol. Also, it would be better to use scientific terms, such as 'emulsification'. More detail could be given about how emulsification helps to increase the rate of lipase action by increasing the surface area of the fat.

Now try this

1. The picture shows the limb of a fossil of an early mammal that lived many millions of years ago.

   Discuss how useful fossils such as this are in providing evidence for the evolution of mammals. **(6 marks)**

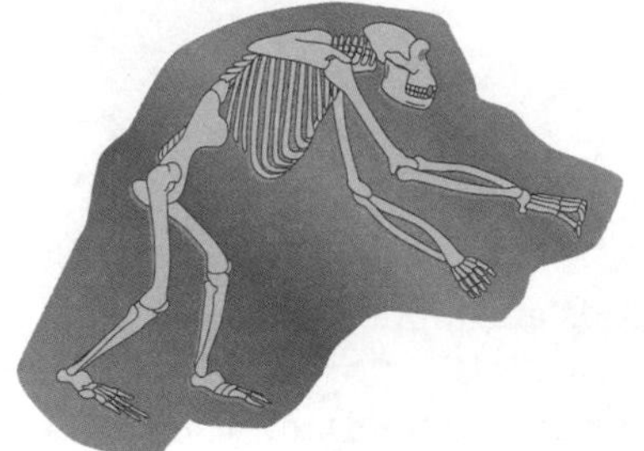

A discussion like this needs to present evidence and draw a conclusion. You need to think about the evidence that fossils can provide for mammalian evolution, but also any limitations with the evidence.

Biology extended writing 5

Worked example

Substances called stanol esters are made from substances found in plants. These stanol esters are often added to food as there is some evidence that they keep us healthy.

Explain how scientists might use evidence from scientific studies to decide if plant stanol esters have an effect on human health. **(6 marks)**

Sample answer 1

Plant stanols are the compounds that are added to some margarines. The people that make them say that they reduce cholesterol. So, you'd give people the margarine to eat and see if they had less cholesterol. Cholesterol is bad for you because it clogs up your arteries.

This is a basic answer. The answer clearly states what plant stanol esters are believed to do and why this is important. There are some ideas of what studies might be carried out to find out if they worked, but there is no evaluation of the evidence.

Sample answer 2

Plant stanol esters are added to spreads. The theory is that they lower cholesterol levels in the blood. To work out if this true, scientists will look at evidence from a proper study. A well-run study should have two groups of people – one group who will carry on as normal (the control group) and a second group who replace their usual spread with one that contains plant stanols. It would be best if the people in the groups matched each other by age and gender. The scientists would measure the level of cholesterol in the blood of the people in each group over a period of a few weeks to see if any changes did happen. Many food companies claim that their spreads are healthy. Scientists would avoid using data provided by these companies, as they might be biased.

This is an excellent answer. It clearly states the way a study would have to be set up so that it could provide the right kind of evidence. It also mentions the sort of studies that should be avoided, because the people setting up the study have a commercial interest in the results.

Now try this

1. Explain how the small intestine is adapted to ensure that carbohydrates and proteins in the food are absorbed into the blood. **(6 marks)**

Structure of the atom

Atoms have protons and neutrons in a central nucleus, with electrons in shells (or energy levels) around the nucleus. The nucleus is very small compared with the overall size of the atom. Atoms of a given element all have the same number of protons in their nuclei. It is the number of protons in the nucleus that determines which element the atom is part of. Atoms have the same number of protons and electrons. The + charges on the protons balance the – charges on the electrons, so the atom is neutral overall.

Particle	Relative mass	Relative charge
proton	1	+1
neutron	1	0
electron	$\frac{1}{1840}$	–1

The masses and charges on subatomic particles are very small, so we only talk about the masses and charges relative to each other. The mass of an electron is so small that we don't bother to account for it in calculations.

Isotopes

The atomic number of an element is the number of protons in the nucleus. The mass number is the total number of protons and neutrons in the nucleus. The number of neutrons = mass number – atomic number.

Isotopes are atoms of the same elements with different numbers of neutrons. This means that some relative atomic masses are not whole numbers.

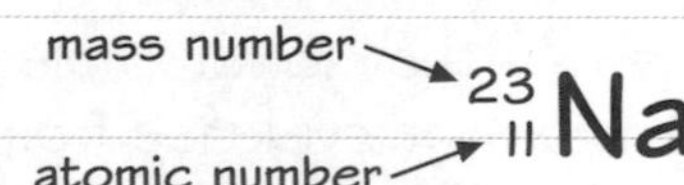

The masses of atoms are so small that we do not talk about their mass in kilograms. Instead we compare the mass of atoms with the mass of carbon-12. This is called the relative atomic mass. Relative atomic mass does not have units.

Worked example

Gallium has two isotopes: $^{69}_{31}$Ga and $^{71}_{31}$Ga (sometimes written as gallium-69 and gallium-71). The relative abundance of $^{69}_{31}$Ga is 60%. Calculate the relative atomic mass of gallium. Give your answer to 1 decimal place.

Relative abundance of $^{71}_{31}$Ga = 100% – 60%
= 40%

$$\text{relative atomic mass} = \frac{(69 \times 60) + (71 \times 40)}{100\%}$$

$$= \frac{4140 + 2840}{100}$$

$$= 69.8$$

Multiply the mass number of each isotope by its relative abundance, and then add them all together and divide by 100.

Now try this

1. The relative atomic mass of copper is 63.5. State the reason why the relative atomic mass of copper is not a whole number. **(1 mark)**

2. Antimony has two isotopes: $^{121}_{51}$Sb (antimony-121, relative abundance 57%) and $^{123}_{51}$Sb (antimony-123). Calculate the relative atomic mass of antimony. Give your answer to 1 decimal place. **(2 marks)**

The modern periodic table

The elements in the modern periodic table are arranged in order of their atomic numbers.

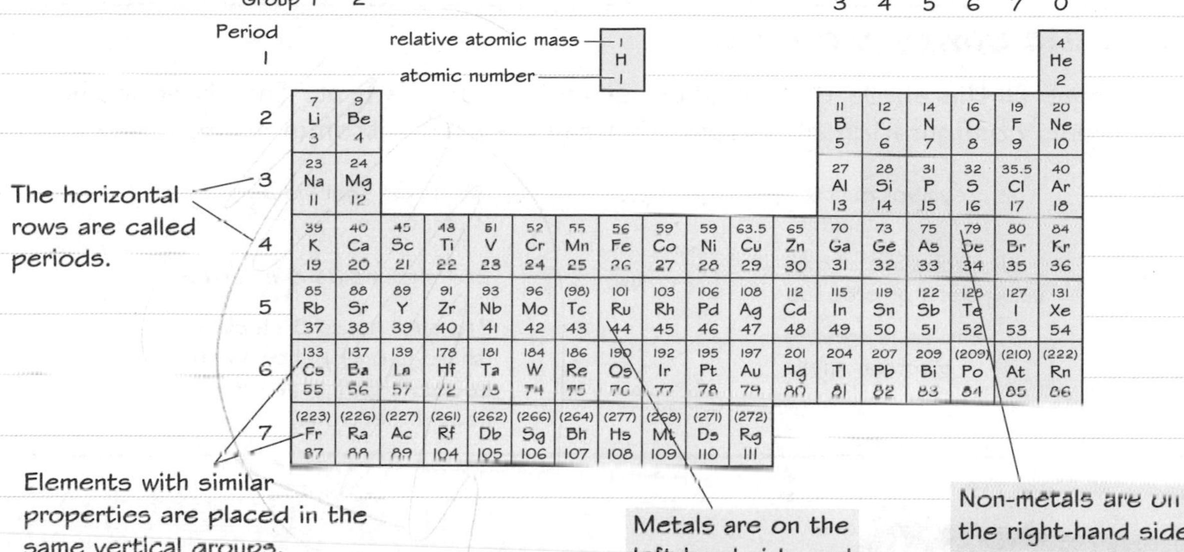
Group 1 2 3 4 5 6 7 0
Period
relative atomic mass
atomic number
The horizontal rows are called periods.
Elements with similar properties are placed in the same vertical groups.
Metals are on the left-hand side and in the centre.
Non-metals are on the right-hand side.

Period	1	2											3	4	5	6	7	0
1								1 H 1										4 He 2
2	7 Li 3	9 Be 4											11 B 5	12 C 6	14 N 7	16 O 8	19 F 9	20 Ne 10
3	23 Na 11	24 Mg 12											27 Al 13	28 Si 14	31 P 15	32 S 16	35.5 Cl 17	40 Ar 18
4	39 K 19	40 Ca 20	45 Sc 21	48 Ti 22	51 V 23	52 Cr 24	55 Mn 25	56 Fe 26	59 Co 27	59 Ni 28	63.5 Cu 29	65 Zn 30	70 Ga 31	73 Ge 32	75 As 33	79 Se 34	80 Br 35	84 Kr 36
5	85 Rb 37	88 Sr 38	89 Y 39	91 Zr 40	93 Nb 41	96 Mo 42	(98) Tc 43	101 Ru 44	103 Rh 45	106 Pd 46	108 Ag 47	112 Cd 48	115 In 49	119 Sn 50	122 Sb 51	128 Te 52	127 I 53	131 Xe 54
6	133 Cs 55	137 Ba 56	139 La 57	178 Hf 72	181 Ta 73	184 W 74	186 Re 75	190 Os 76	192 Ir 77	195 Pt 78	197 Au 79	201 Hg 80	204 Tl 81	207 Pb 82	209 Bi 83	(209) Po 84	(210) At 85	(222) Rn 86
7	(223) Fr 87	(226) Ra 88	(227) Ac 89	(261) Rf 104	(262) Db 105	(266) Sg 106	(264) Bh 107	(277) Hs 108	(268) Mt 109	(271) Ds 110	(272) Rg 111							

Mendeleev

The periodic table as we know it was first made by Mendeleev, who arranged all the elements known at the time into a table.

Mendeleev put the elements in order of the relative atomic mass.

He checked the properties of the elements and their compounds.

He swapped the places of some elements so that elements with similar properties lined up.

He left gaps where he thought there were other elements, and predicted their properties.

When these elements were discovered, Mendeleev's predictions fitted the properties very well!

Now try this

target D-C

1. Give the symbol for a non-metal in period 3 of the periodic table. **(1 mark)**

target B-C

2. Look at the elements with atomic numbers 52 and 53 (tellurium and iodine).
   (a) Explain how Mendeleev could have arranged these two elements initially. **(2 marks)**
   (b) Suggest why Mendeleev would have then swapped them over. (Scientists did not know about protons and atomic numbers when Mendeleev developed the periodic table). **(2 marks)**

Electron shells

The electrons in an atom are arranged in electron shells, or energy levels around the nucleus.

Electronic configurations

The electrons fill the shells that are closest to the nucleus first. The diagram shows the electronic configuration of sodium. This can also be written in numbers: 2.8.1

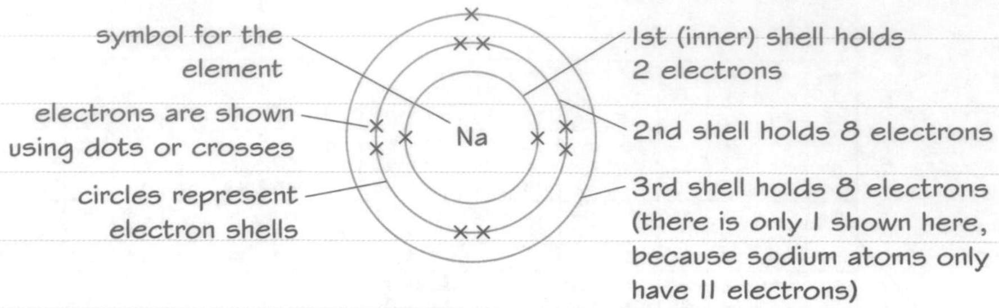

symbol for the element

electrons are shown using dots or crosses

circles represent electron shells

Na

1st (inner) shell holds 2 electrons

2nd shell holds 8 electrons

3rd shell holds 8 electrons (there is only 1 shown here, because sodium atoms only have 11 electrons)

Worked example

The atomic number of potassium is 19. Explain how its electrons are arranged.

An atom has the same number of protons and electrons, so a potassium atom has 19 electrons. The inner shell fills up first, and this shell is full when it has 2 electrons. The second shell has 8 electrons, and this is also full. This leaves 9 electrons. The third shell can hold 8 electrons, so that is full as well. The last shell has 1 electron in it. We write this as 2.8.8.1.

You do not need to learn the electronic configurations of any of the elements. But if you are told the atomic number of an element you need to be able to work out the electronic configuration.

Electrons and groups

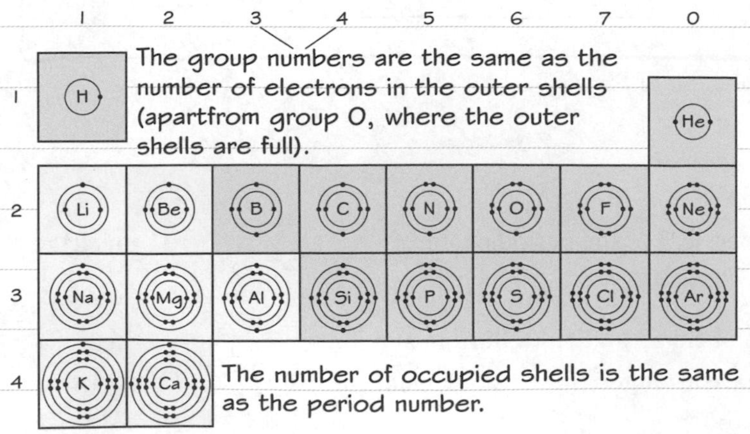

	1	2	3	4	5	6	7	0
1	H							He
2	Li	Be	B	C	N	O	F	Ne
3	Na	Mg	Al	Si	P	S	Cl	Ar
4	K	Ca						

The group numbers are the same as the number of electrons in the outer shells (apartfrom group 0, where the outer shells are full).

The number of occupied shells is the same as the period number.

Now try this

target D-C

1. Write the electronic configurations of these elements:
   (a) lithium (Li, atomic number = 3) **(1 mark)**
   (b) oxygen (O, atomic number = 8) **(1 mark)**
   (c) chlorine (Cl, atomic number = 17) **(1 mark)**

target C-B

2. The atomic number of an element is 12. State what this tells you about the structure of the atoms in this element and its position in the periodic table. **(4 marks)**

Ions

Atoms and ions

An atom has the same number of protons and electrons. It has no overall charge, because the positive charges on the protons are balanced by the negative charges on the electrons.

An atom can become an ion by gaining or losing one or more electrons. An ion has a charge because it no longer has the same number of protons and electrons. We show that the atom has become an ion by adding a + or – sign to the symbol.

Cations

Metal atoms lose electrons to form positively charged cations.

- Metals in group 1 lose 1 electron to form ions with a 1+ charge.
- Metals in group 2 lose 2 electrons to form ions with a 2+ charge.

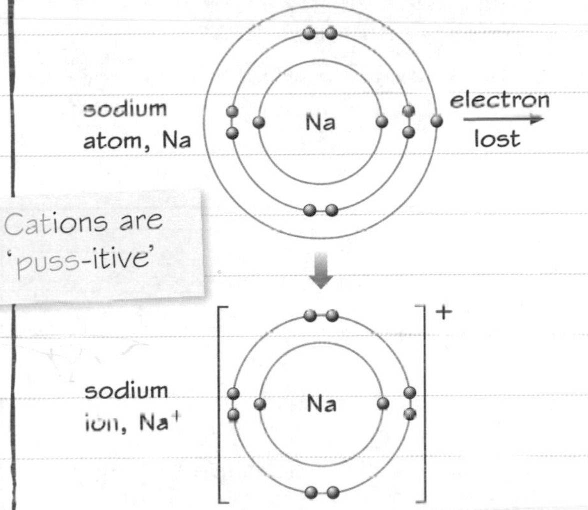

Cations are 'puss-itive'

Anions

Non-metal atoms (or groups of atoms) gain electrons to form negatively charged anions.

- Elements in group 6 gain 2 electrons to form ions with a 2– charge.
- Elements in group 7 gain 1 electron to form ions with a 1– charge.

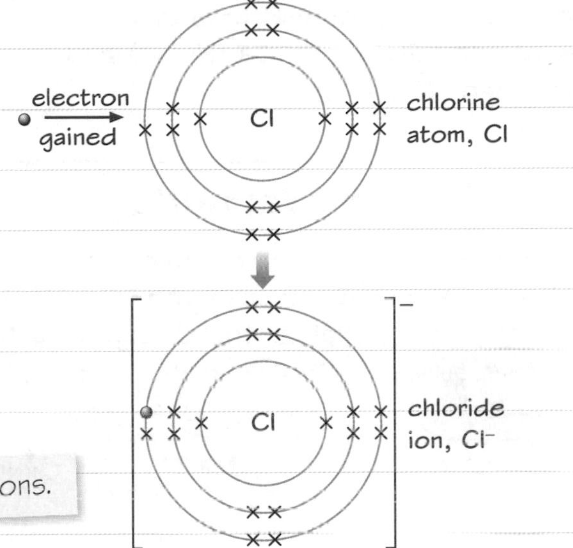

Non-metal atoms form Negative ions.

The electrons in metals and non-metals are exactly the same. We just use dots and crosses to help us to tell them apart on diagrams.

Now try this

1. Write down the ions formed by the following elements.
   (a) calcium (Ca) **(1 mark)**
   (b) fluorine (F) **(1 mark)**
   (c) potassium (K) **(1 mark)**
   (d) sulfur (S) **(1 mark)**
2. Draw labelled diagrams similar to the ones above to show how electrons are gained and lost in the formation of lithium fluoride. **(3 marks)**

EXAM ALERT!

Remember the metal atoms always lose electrons to form positive ions, and non-metal atoms gain electrons to form negative ions. On a recent exam question, over a third of students got no marks at all on a question about this.

Students have struggled with exam questions similar to this - **be prepared!** ResultsPlus

Ionic compounds

When sodium and chlorine react, sodium atoms lose electrons and chlorine atoms gain electrons. The sodium and chloride ions have opposite charges, and they attract each other. This creates ionic bonds between the ions and a compound is formed. An ionic compound formed between a metal and a non-metal is called a salt.

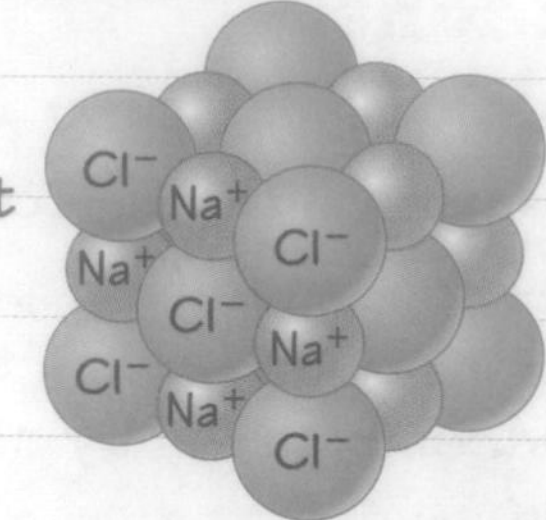

The lattice structure of sodium chloride.

Compound ions

Some groups of atoms can form ions. These are called compound ions.

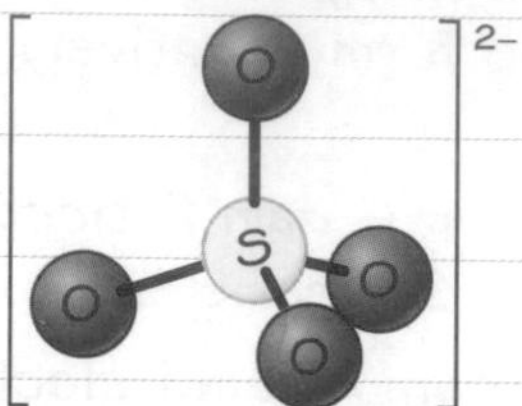

sulphate ion SO_4^{2-}

Some other compound ions are: hydroxide (OH^-), nitrate (NO_3^-) and carbonate (CO_3^{2+}).

The name of a compound tells you about the elements in it – so sodium chloride contains sodium and chlorine in the form of ions.

If the name of a compound ends in '**-ate**' it shows that oxygen is part of the compound as well. For example calcium carbonate contains calcium, carbon and oxygen.

A name ending in '**-ide**' shows that no extra oxygen is present. For example, copper sulfide contains only copper and sulfur. (Sodium hydroxide contains oxygen, but the 'ox' part of the name tells you that.)

Finding the formula

You need to be able to work out the formula of a compound if you are given the ions and their charges. The key thing is to remember that all the charges must cancel out.

calcium sulfide: Ca^{2+} S^{2-} → CaS

sodium oxide: Na^+ O^{2-} → Na_2O

copper nitrate: Cu^{2+} NO_3^- → $Cu(NO_3)_2$

aluminium oxide: Al^{3+} O^{2-} → Al_2O_3

The nitrate ion is a compound ion. Use brackets if you need to put a number with a compound ion (e.g. magnesium hydroxide is $Mg(OH)_2$). If there is only one compound ion in a formula you do not need the brackets (e.g calcium carbonate is $CaCO_3$).

Easy method:

- Swap the numbers that give the charge for each ion and write them after the relevant symbol.
- If the number is 1, don't write it in the formula.
- If both the numbers are the same, don't write them in the formula.

Now try this

1. Write down the formulae of the compounds formed from:
   (a) calcium ions (Ca^{2+}) and sulfide ions (S^{2-}) **(1 mark)**
   (b) magnesium ions (Mg^{2+}) and nitrate ions (NO_3^-) **(1 mark)**
   (c) potassium ions (K^+) and sulfate ions (SO_4^{2-}). **(1 mark)**

target B-A*

2. Work out the formula of the compound formed between magnesium (Mg, atomic number 12) and fluorine (F, atomic number 9). Explain how you worked out your answer. **(5 marks)**

Properties of ionic compounds

Melting and boiling points

Ionic substances have high melting points and high boiling points.
This is because the ions are held together in the lattice by very strong electrostatic forces. A lot of energy is needed to break these bonds.
Ionic substances are solids at room temperature and have to be heated strongly to make them melt.

EXAM ALERT!

In a recent examination, nearly half the students scored no marks at all on a question about the high melting point of sodium chloride. Remember, ions are held together by strong electrostatic forces so it takes a lot of energy (high temperatures) to break up the lattice and form a liquid.

Students have struggled with exam questions similar to this - **be prepared!** ResultsPlus

Conducting electricity

A substance only conducts electricity if it contains charged particles that are free to move. Ionic substances do not conduct electricity when they are solid because the ions are held together in a lattice. They do conduct electricity when they are molten or dissolved in water ('in aqueous solution') because the charged particles can move around.

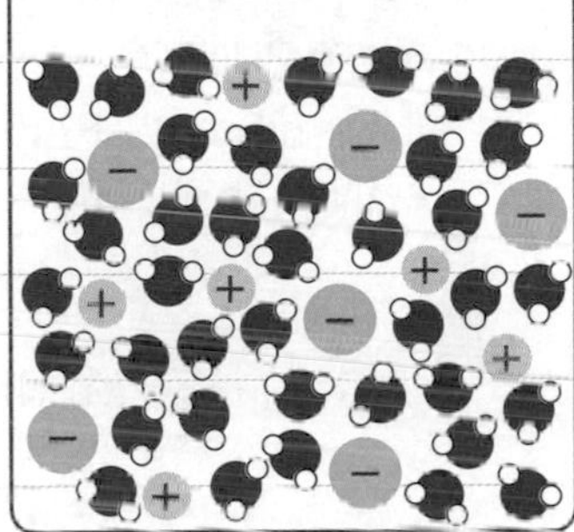

The ions can move around in an aqueous solution.

Solubility rules

Many ionic substances dissolve in water. You need to learn the rules that help you to work out whether a particular substance is soluble.

If you can remember this rule it will help you to remember which carbonates and hydroxides *do* dissolve.

Soluble in water	Exceptions (insoluble in water)
all common salts of • sodium • potassium • ammonium	
all nitrates	
most chlorides	silver chloride, lead chloride
most sulfates	lead sulfate, barium sulfate, calcium sulfate

Insoluble in water	Exceptions (soluble in water)
most carbonates	sodium carbonate potassium carbonate ammonium carbonate
most hydroxides	sodium hydroxide potassium hydroxide ammonium hydroxide

There are no short cuts here – you just need to *learn* all of these rules.

Now try this

1. Explain whether each of these substances conducts electricity:
   (a) molten barium sulfate **(2 marks)**
   (b) copper sulfate powder **(2 marks)**
   (c) sodium chloride solution. **(2 marks)**

2. State whether each of these substances is soluble or insoluble in water:
   (a) barium sulfate **(1 mark)**
   (b) cobalt sulfate **(1 mark)**
   (c) magnesium hydroxide **(1 mark)**

Precipitates

If you mix solutions of two soluble salts, the ions in the mixture can combine to form new salts. If one of the new combinations produces an insoluble salt, it will appear as a precipitate. You can use the solubility rules to work out if a precipitate will form.

Predicting precipitates

Worked example

Silver nitrate solution is mixed with potassium chloride solution. Explain why a precipitate will form and write equations for the reaction.

Silver chloride is insoluble, so when silver and chloride ions come into contact they will form a precipitate.

silver nitrate + potassium chloride → silver chloride + potassium nitrate

$AgNO_3(aq)$ + $KCl(aq)$ → $AgCl(s)$ + $KNO_3(aq)$

The two substances on the left are solutions. The precipitate of silver chloride is a solid.

Preparing salts

Pure samples of insoluble salts can be prepared using precipitation reactions.

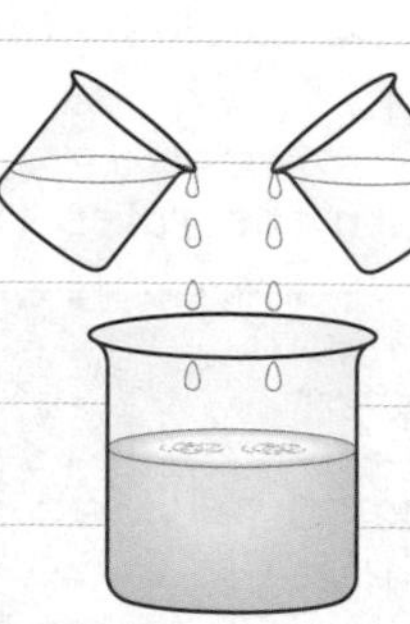

Mix solutions of two substances that will form the insoluble salt.

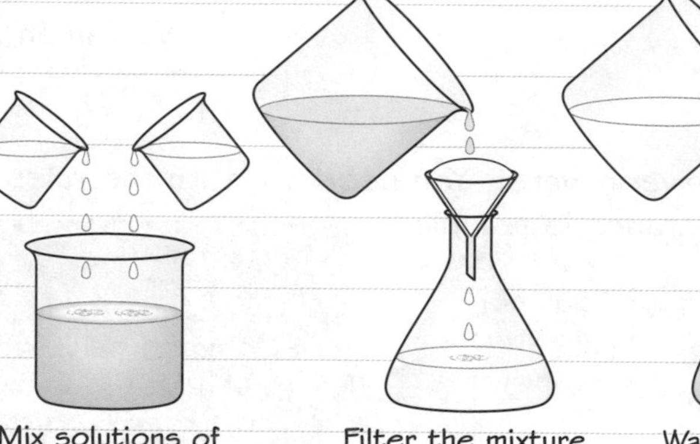

Filter the mixture. The insoluble salt will be trapped in the filter paper.

Wash the salt with pure water.

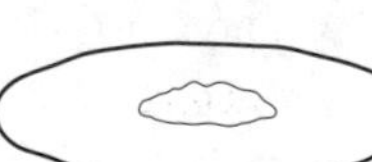

Leave the salt to dry on the filter paper. It could be dried in an oven.

Barium sulfate

Barium sulfate is an insoluble salt. It is given as a 'barium meal' to patients who are going to have their stomachs or intestines X-rayed because:

- barium sulfate is opaque to X-rays, so it shows up well on the X-ray picture
- it is safe to use (barium salts are toxic, but because barium sulfate is insoluble barium does not get into the patient's blood).

Now try this

1. Explain what will happen when these solutions are mixed:
   (a) sodium carbonate and calcium nitrate **(3 marks)**
   (b) potassium nitrate and ammonium carbonate. **(2 marks)**
2. Suggest two solutions you could mix to obtain a precipitate of calcium hydroxide. **(2 marks)**

Ion tests

Some of the ions in salts can be identified using tests in the laboratory.

Flame tests

Some metal ions can be identified using flame tests. A damp splint is placed in the solid and then held in a Bunsen burner flame. The colour of the flame identifies the ions.

Metal ion	Flame test colour
sodium (Na^+)	yellow
potassium (K^+)	lilac
calcium (Ca^{2+})	red
copper (Cu^{2+})	green–blue

Spectroscopy

Worked example

Describe how a spectroscope was used to discover new elements.

Very small amounts of some elements can be detected in flame tests by using a spectroscope. The elements rubidium and caesium were discovered when scientists spotted colours in parts of the spectrum they had not seen before.

Carbonate ions (CO_3^{2-})

Test:

- Add dilute acid to the substance.
- Test the gas given off with limewater.

If the substance contains carbonate ions:

- it will fizz or bubble as a gas is given off
- the gas will turn limewater milky, showing the gas is carbon dioxide.

Sulfate ions (SO_4^{2-})

Test:

- Add a few drops of dilute hydrochloric acid.
- Shake the mixture.
- Add a few drops of barium chloride solution.

If the substance contains sulfate ions:

- a white precipitate of barium sulfate forms.

Chloride ions (Cl^-)

Test:

- Add a few drops of dilute nitric acid to the solution.
- Shake the mixture.
- Add a few drops of silver nitrate solution.

If the substance contains chloride ions:

- a white precipitate of silver chloride forms.

EXAM ALERT!

If you are asked to describe the test for a substance, you need to say how to carry out the test *and* what you would see if the substance is the one you expect.

Students have struggled with exam questions similar to this - **be prepared!** ResultsPlus

Now try this

1. A white solid produces a red flame in a flame test. The substance fizzes when dilute acid is added to it.

(a) Suggest what compound this might be. **(1 mark)**

(b) Explain your answer to part **(a)**. **(3 marks)**

(c) Explain what further test you would need to carry out to confirm the identity of the compound. **(2 marks)**

Chemistry extended writing 1

To answer an extended writing question successfully you need to:
- use your scientific knowledge to answer the question
- organise your answer so that it is logical and well ordered
- use full sentences in your writing and make sure that your spelling, punctuation and grammar are correct.

Worked example

Explain how bonds form when sodium reacts with chlorine, and how the properties of sodium chloride depend on what these bonds are like. **(6 marks)**

Sample answer 1

Sodium chloride is an ionic compound. It conducts electricity when it is molten or in solution. It has a high melting point and boiling point.

This is a basic answer. The information given is correct, but the answer does not explain how the properties depend on the bonds, or how the bonds are formed. It is important to answer everything the question asks.

Sample answer 2

Sodium is an alkali metal and its atoms have 1 electron in the outer shell. When sodium reacts with chlorine, each atom of sodium loses its outer electron to form an ion with a positive charge. Chlorine atoms have 7 electrons in their outer shell, and when they react with sodium they gain an electron to form ions with a negative charge. The positive and negative charges attract each other. This is an ionic bond, and ionic bonds hold all the ions together in a lattice structure.

The ionic bonds are very strong, so it takes a lot of energy to break them, which means that sodium chloride has high melting and boiling points. The ions can carry an electric current if they can move freely, because they have electrical charges. The ions can move and sodium chloride can conduct electricity if it is dissolved in water or is melted.

This is an excellent answer that contains all the information asked for in the question. It uses correct scientific terms such as lattice structure, atom and ion, and it is organised in a sensible order.

Now try this

1. Magnesium has an atomic number of 12. Chlorine has an atomic number of 17. Explain what happens when magnesium reacts with chlorine. **(6 marks)**

Covalent bonds

Molecules consist of two or more atoms chemically joined together. The atoms are held together by covalent bonds. A covalent bond is a pair of electrons that is shared between two atoms.

In all kinds of bonding, atoms lose, gain or share electrons to get a full outer shell. Covalent bonding takes place between non-metal atoms. Non-metal atoms need one or more electrons to fill their outer shell, and they do this by sharing the electrons.

Dot and cross diagrams

We can show the formation of simple, molecular covalent substances using dot and cross diagrams.

The electrons in the atoms are all the same. We use dots and crosses to help us see which electrons came from the different atoms.

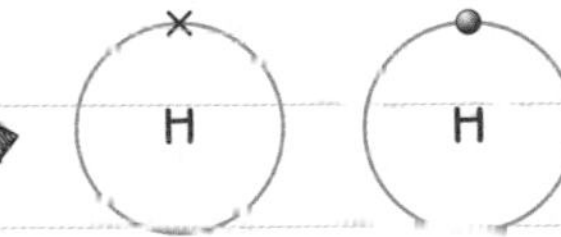

Hydrogen atoms each have one electron. The first shell can hold two electrons, so each atom needs one more electron to have a full shell.

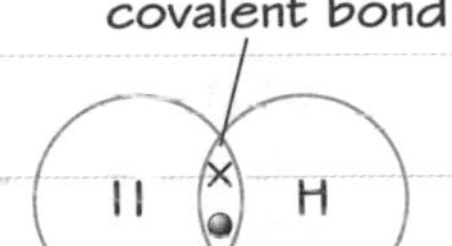

A hydrogen molecule has two hydrogen atoms sharing their electrons. Both atoms now have a full outer shell.

Worked example

Draw dot and cross diagrams to show the covalent bonds in:

(a) hydrogen chloride (HCl) **(b)** water (H_2O) **(c)** carbon dioxide (CO_2).

a)

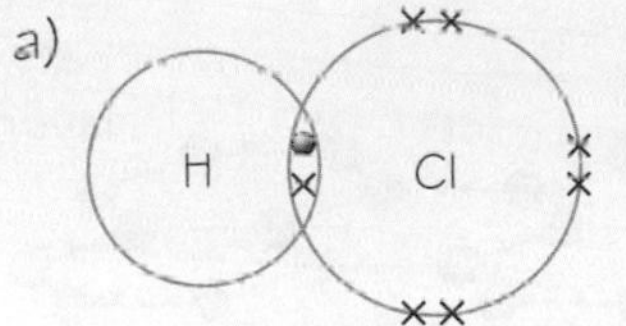

b)

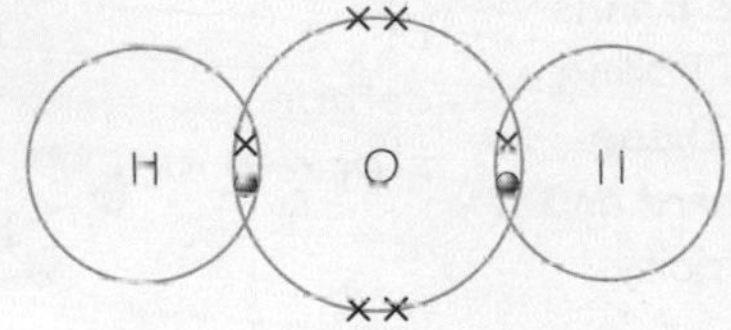

c)

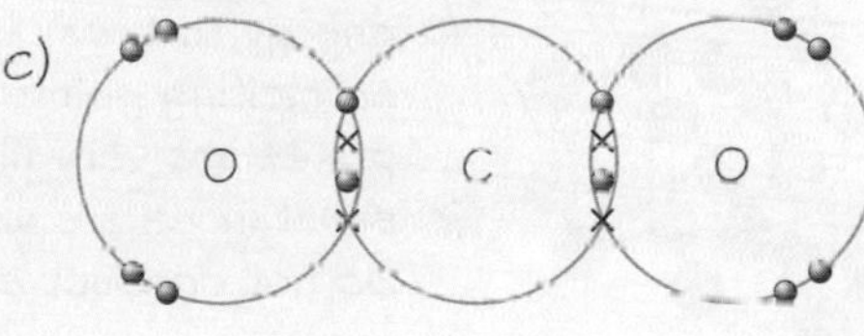

The diagrams only show the outer electron shells for each atom, as these are the ones involved in the bonds. Look back at page 40 to remind yourself about the arrangement of electrons in atoms.

Now try this

1. The electronic configuration of carbon is 2.4.
   - **(a)** State how many electrons a carbon atom needs to fill its outer shell. **(1 mark)**
   - **(b)** Explain why the formula of methane is CH_4. **(3 marks)**
   - **(c)** Draw a dot and cross diagram to show the covalent bonds in methane. **(2 marks)**

target B-A

2. The electronic configuration of oxygen is 2.6.
   - **(a)** Explain why there is a double covalent bond in an oxygen molecule (O_2). **(2 marks)**
   - **(b)** Draw a dot and cross diagram to show the covalent double bond in an oxygen molecule. **(2 marks)**

Covalent substances

Substances with covalent bonds can form small molecules or giant structures. The two types of substances have different properties.

Simple molecular covalent substances

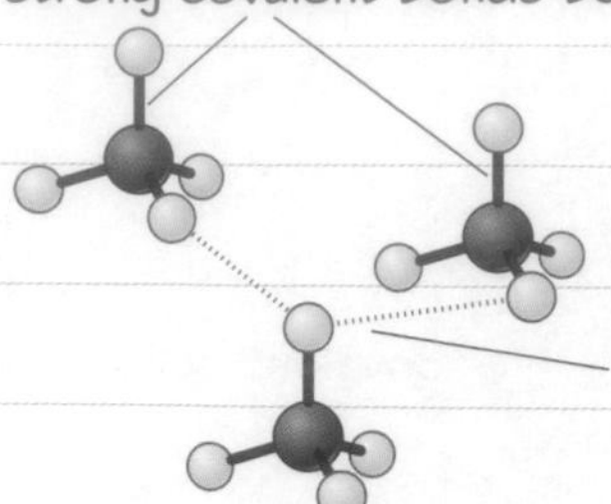

There are only weak forces between the molecules.These substances have low melting points and low boiling points. These substances are liquids or gases at room temperature, or solids with low melting points, such as sugar, wax or iodine.

There are no charged particles, so simple molecular covalent substances do not conduct electricity.

EXAM ALERT!

Remember which bonds are strong and which are weak. Nearly half of students got no marks on a recent question about this.

Students have struggled with this topic in recent exams - **be prepared!** ResultsPlus

Giant molecular covalent substances

Diamond is a giant molecular covalent substance and is made of billions of atoms all joined.

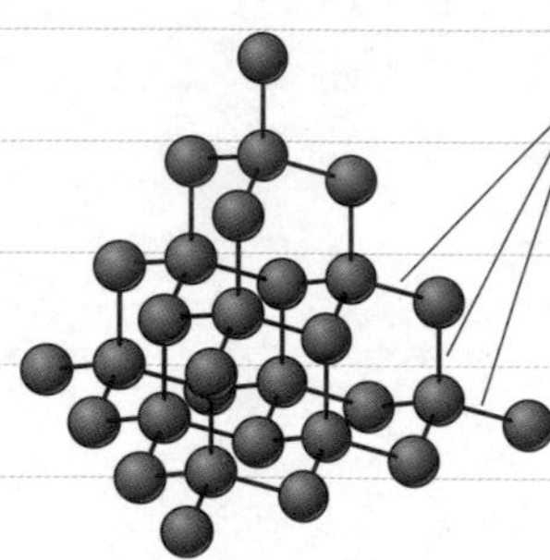

The covalent bonds in giant molecular covalent substances are strong. It takes a lot of energy to break these bonds so melting points and boiling points are very high. These substances are very hard and do not conduct electricity.

Graphite is also a giant molecular covalent substance, so it has a high melting point, but some of its properties are different to diamond. It does conduct electricity, and is very soft.

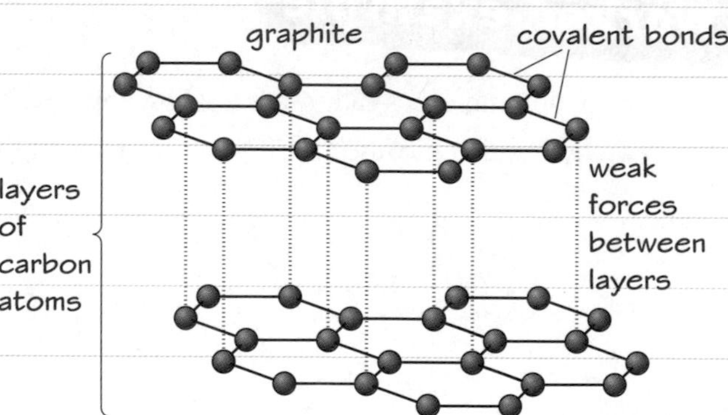

Diamond is used for cutting tools because it is very hard. Graphite is used as a lubricant because the forces between the layers are weak and the layers can slide over each other easily. In graphite, there is one electron from each carbon atom that can move along the layers, so graphite conducts electricity and can be used to make electrodes.

Now try this

1. Sand (silicon dioxide) is made of silicon and oxygen atoms joined together in a giant covalent structure. Explain why:
   (a) grains of sand are hard **(1 mark)**
   (b) sand does not conduct electricity **(1 mark)**
   (c) sand is a solid at room temperature. **(2 marks)**
2. Carbon dioxide is made from carbon and oxygen atoms joined together by covalent bonds. Explain why carbon dioxide is a gas at room temperature when silicon dioxide is a solid. **(3 marks)**

Miscible or immiscible?

Miscible liquids mix completely with each other. Immiscible liquids do not mix completely.

Separating immiscible liquids

Worked example

The diagram shows a separating funnel. Explain how this can be used to separate two immiscible liquids.

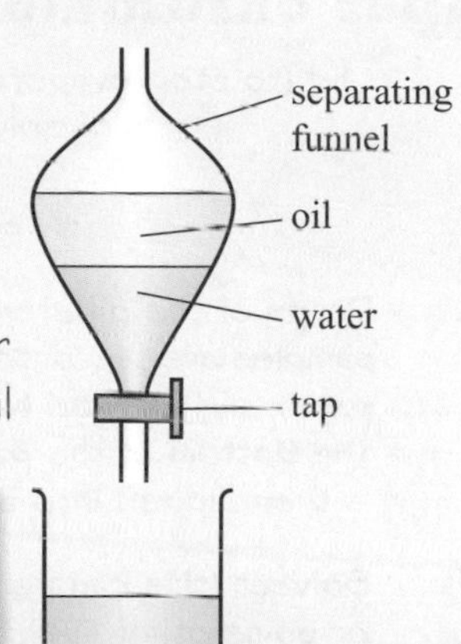

Place the mixture in the funnel and put a beaker underneath.

Open the tap until most of the lower liquid has run into the beaker.

Put a clean beaker under the funnel and run the last bit of the lower liquid and a little bit from the top liquid into the beaker. Place a final clean beaker under the tap and run off all of the second liquid.

The liquid with the lowest density is at the top. In this example, you can see that oil is less dense than water.

Separating miscible liquids

Miscible liquids can be separated using fractional distillation.

- The mixture is heated, and the vapours pass into a fractionating column.
- As the gases cool down, the ones with the highest boiling points condense first.
- Different liquids are collected at different places in the column.

Fractional distillation only works if the two liquids in the mixture have different boiling points.

Distillation of liquid air

Nitrogen and oxygen are obtained from air by fractional distillation. The air has to be turned to a liquid first by cooling it.

Remember: −190 °C is colder than −185 °C.

The top of the column is colder. At −190 °C, the nitrogen is still a gas but any of the oxygen reaching this part of the tower is likely to condense and run to the bottom of the column.

−190 °C

nitrogen gas is collected

−185 °C is above the boiling point of nitrogen and so the nitrogen boils and evaporates.

fractionating column

cooling unit

air in

−185 °C

−185 °C is below the boiling point of oxygen and a lot of the oxygen stays as a liquid (although some evaporates).

liquid air at −200 °C

liquid oxygen piped out

−200 °C is below the boiling points of both nitrogen and oxygen and so they are both liquids.

Now try this

1. Salad dressing is a mixture of oil and vinegar. Oil and vinegar are immiscible liquids. Explain why a bottle of salad dressing needs to be shaken before it is used. **(2 marks)**
2. Explain why the following temperatures are used in a fractionating column used to separate nitrogen and oxygen from air:
   (a) −200 °C for the air going into the column **(2 marks)**
   (b) −185 °C at the bottom of the column. **(2 marks)**

Chromatography

Most inks, paints, dyes and food colourings are mixtures of different coloured compounds. Chromatography can be used to separate the different substances in these mixtures.

> If you are asked to draw the apparatus for chromatography, make sure the line for the solvent is above the bottom edge of the paper but below the samples. If the samples are dipped into the solvent they will just dissolve into it.

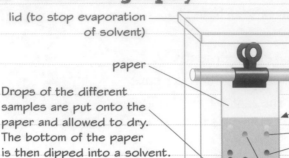

Paper chromatography

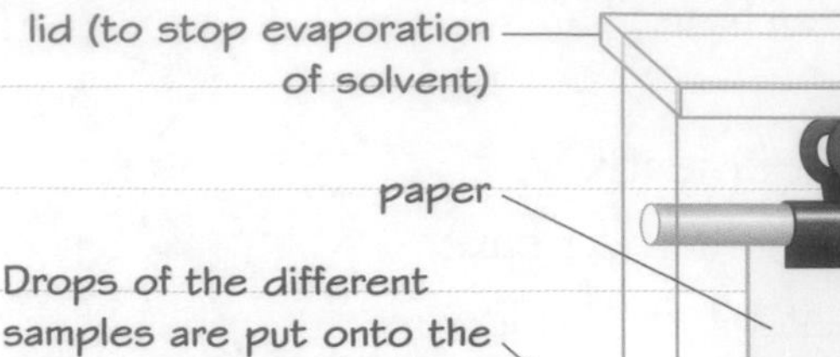
lid (to stop evaporation of solvent)
paper
Drops of the different samples are put onto the paper and allowed to dry. The bottom of the paper is then dipped into a solvent.
Solvent (this can be water or some other liquid that the samples will dissolve in)
X Y Z
Solvent front (the solvent has reached this level)
The different compounds in a sample dissolve to different extents in the solvent.
More soluble compounds are carried up the paper faster than less soluble ones, so the compounds separate out.

R_f values

The R_f value of a compound is always the same as long as the chromatography is carried out in the same way, under the same conditions. R_f values can be used to identify compounds from a chromatogram.

$$R_f = \frac{\text{distance moved by compound}}{\text{distance moved by solvent}}$$

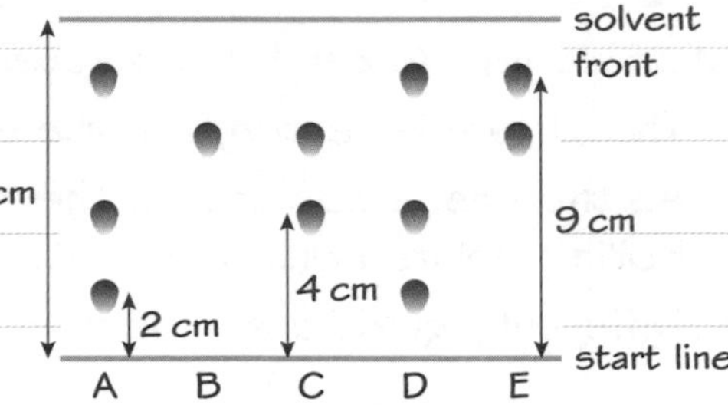
solvent front
10 cm
9 cm
4 cm
2 cm
start line
A B C D E

> R_f is one distance divided by another, so it has no units.

Worked example

Calculate the R_f value for the lowest spot for sample A on the chromatogram above.

$$R_f = \frac{2\text{ cm}}{10\text{ cm}} = 0.2$$

> A compound never gets as far up the paper as the solvent, so the R_f value must always be less than 1.

Chromatography can be used...

- in the food industry, to check which colourings are used in food
- in forensic science, to analyse DNA samples or paints and inks from crime scenes
- by museums to analyse paints, to help them to restore old paintings or detect forgeries.

Now try this

All of these questions relate to the chromatogram shown in the middle of the page.

target
D-C

1. Calculate the R_f value for the lowest spot for sample C. **(2 marks)**
2. Which spot (or spots) have an R_f value of 0.9? **(1 mark)**

target
B-C

3. Sample A contains 3 different substances. Explain what the chromatogram tells you about the other four substances tested. **(4 marks)**

Chemical classification 1

Elements and compounds can be classified into different types.

Ionic substances

Bonding:
- atoms gain or lose electrons to form ions
- strong bonds between ions

Electricity:
- conduct electricity when molten or dissolved in water, because the charged ions can move around
- do not conduct when solid, as the charged particles cannot move

sodium chloride

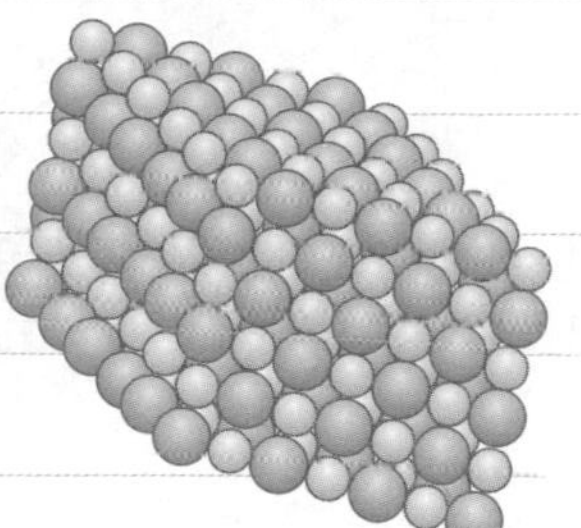

Melting and boiling points:
- high, because there are strong forces holding the ions together
- ionic substances are solids at room temperature

Solubility:
- many dissolve in water

Simple molecular covalent substances

Bonding:
- atoms share electrons to get a full outer shell
- strong bonds between atoms in molecules
- weak forces between separate molecules

Electricity:
- do not conduct electricity, as there are no charged particles that can move around

water, oxygen

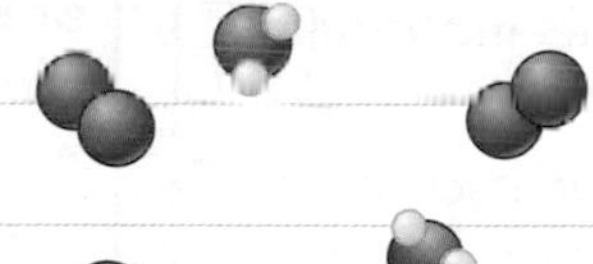

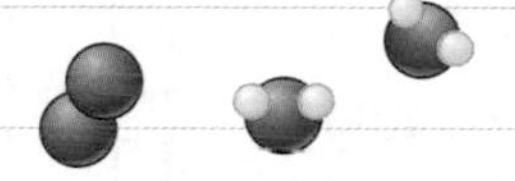

Melting and boiling points:
- low, because there are only weak forces of attraction between the molecules
- most simple molecular covalent substances are liquids or gases at room temperature

Solubility:
- some dissolve in water

Giant molecular covalent substances

Bonding:
- atoms share electrons to get a full outer shell
- strong bonds between all the atoms in a structure

Electricity:
- do not conduct electricity, because there are no charged particles that can move around (except for graphite, where electrons can move along the layers)

sand (silicon dioxide)

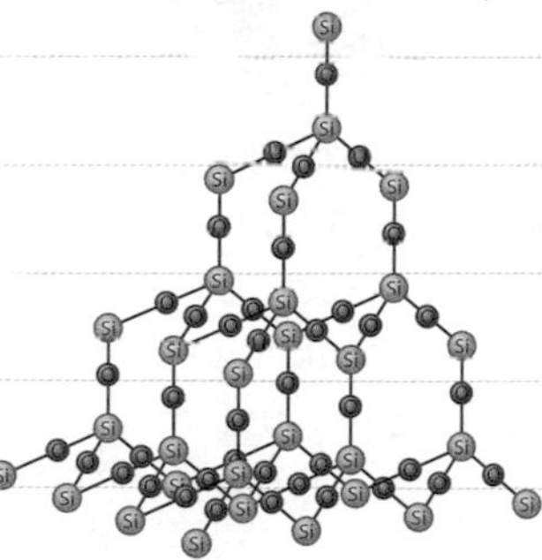

Melting and boiling points:
- high, as there are strong forces holding all the atoms together
- giant molecular covalent substances are solids at room temperature

Solubility:
- insoluble (do not dissolve in water)

Now try this

1. List the types of substances that:
   (a) never conduct electricity **(1 mark)**
   (b) are always solids at room temperature **(1 mark)**
   (c) form structures made from millions of atoms. **(1 mark)**
2. Describe one similarity and two differences between ionic and covalent bonding. **(3 marks)**

Chemical classification 2

Elements and compounds can be classified into different types.

Metals

Bonding:
- positive ions held together by a sea of delocalised electrons
- strength of bonds is stronger in some metals than others

Electricity:
- conduct electricity when solid and liquid, because the delocalised electrons can move between the ions

any metal

Melting and boiling points:
- medium to high
- all metals except mercury are solids at room temperature

Solubility:
- insoluble (do not dissolve in water)

Classifying substances

You can carry out simple tests on substances to help you classify them into ionic substances, metals, simple molecular covalent substances and giant molecular covalent substances.

Melting point:
- If a substance is a liquid or a gas at room temperature it is probably a simple molecular covalent substance.
- If the substance is a solid, put some of it in a crucible and heat it using a Bunsen burner.
- If the substance melts easily it is probably simple molecular covalent. Most metals and giant covalent substances have very high melting points and many giant ionic substances cannot be melted with the Bunsen burner.

Conducting electricity: Use a battery, bulb and some wires to see if the substance conducts electricity.
- If it conducts electricity when it is solid it is a metal (or graphite).
- If it does not conduct when it is solid, try dissolving it in water. If it dissolves and the solution *does* conduct electricity, then it is an ionic substance.

Worked example

How can you use the solubility of a substance to help you to classify it?

I can test the substance to see if it dissolves in water. If it does dissolve it could be an ionic substance, or a simple molecular covalent substance. If it does not dissolve, it does not show that it is *not* one of these substances, as not all ionic or simple molecular covalent substances are soluble.

When you are describing tests, be sure to say what results you may get and what these results mean.

You need to use the results of all the different tests for a substance to help you to decide how to classify it.

Now try this

1. You test a substance that is solid at room temperature and find that it conducts electricity. Explain why this test shows that the substance is a metal and not an ionic substance. **(3 marks)**
2. You have a sample of a substance that dissolves in water.
   (a) Explain how you know that this substance is either an ionic or a simple molecular covalent substance but not a giant molecular covalent substance. **(2 marks)**
   (b) Explain how you would use a battery, bulb and some wire to find out whether it is an ionic substance or a simple molecular covalent substance. **(2 marks)**

Metals and bonding

Metals are malleable and can conduct electricity. To understand why you need to know about the bonding in them.

Metallic bonding

A metal consists of a regular arrangement of positive ions surrounded by a 'sea' of delocalised electrons. These electrons come from the outer shells of the atoms and can move around throughout the metal.

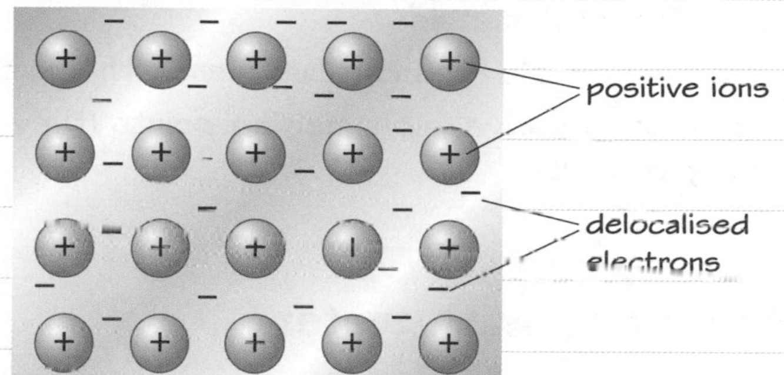

Worked example

Describe how the bonding in metals explains their malleability and ability to conduct electricity.

Metals consist of positive ions surrounded by a 'sea' of delocalised electrons. Metals conduct electricity because the delocalised electrons can move between the metal ions. When there is a potential difference across a piece of metal, the delocalised electrons drift in one direction to form an electric current.

Metals are malleable because the layers of positive ions can slide over each other if a force is applied to the metal.

Malleable materials can be hammered into shape without breaking.

EXAM ALERT!

Over half of students scored no marks in a recent question asking them to explain why metals conduct electricity. Make sure you refer to electrons moving between the ions.

Students have struggled with exam questions similar to this - **be prepared!** ResultsPlus

Transition metals

Transition metals are the metals in the central block of the periodic table. Most metals are transition metals.

Transition metals typically:

- have high melting points
- form coloured compounds.

Be careful not to say that transition metals 'are coloured' or 'form colours'.

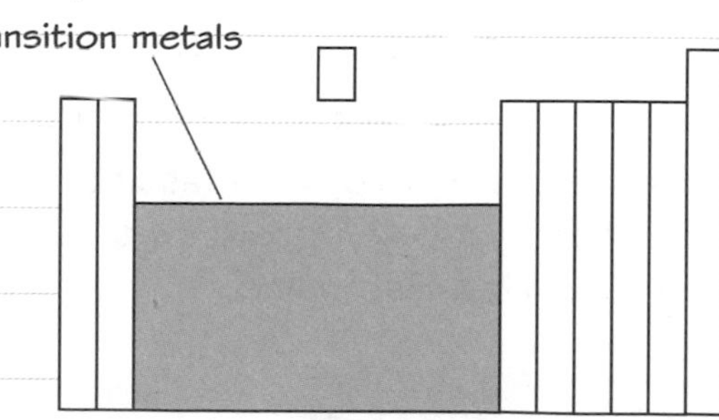

Now try this

1. (a) Describe the structure of metals. **(3 marks)**
   (b) Explain why metals conduct electricity. **(2 marks)**
2. (a) State the meaning of malleable. **(1 mark)**
   (b) Explain why metals are malleable. **(2 marks)**

Alkali metals

Properties

The alkali metals are in group 1 of the periodic table. The alkali metals:

- are soft metals (you can cut them with a knife)
- have comparatively low melting points (they are easy to melt).

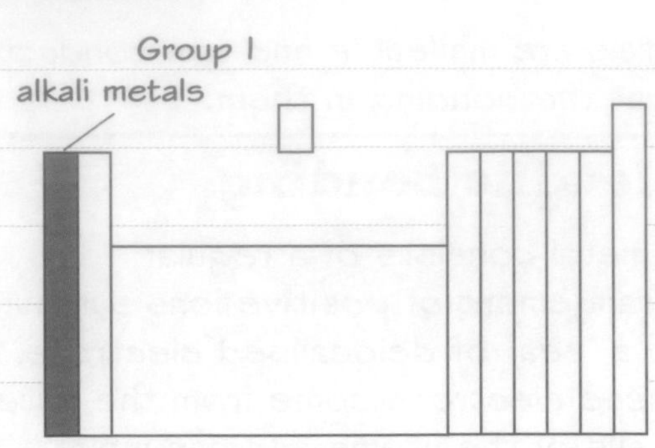

Reactions with water

The alkali metals are all in the same group of the periodic table so they all have similar reactions. They allhave one electron in the outer shell, so when they react, each atom loses one electron to become an ion with a +1 charge.

All the alkali metals react with water to form a metal hydroxide and hydrogen gas. Metal hydroxide solutions are alkaline.

lithium + water → lithium hydroxide + hydrogen

$2Li(s) + 2H_2O(l) \rightarrow 2LiOH(aq) + H_2(g)$

You may be asked to write balanced equations for this reaction, or for the reactions of sodium or potassium. The other hydroxides are NaOH and KOH.

Remember to talk about 'more electron shells' and that the outer electron shell is further from the nucleus for the elements at the bottom of the group.

Reactivity

The reactivity of the alkali metals increases down the group. When they are put in water:

- lithium floats on the surface and fizzes
- sodium melts from the heat produced by the reaction and whizzes around on the surface as a molten ball, and sometimes the hydrogen produced catches fire
- potassium reacts even faster, and the hydrogen produced burns with a lilac flame.

Worked example

Explain why the reactivity of the alkali metals increases as you go down the group.

The elements towards the bottom of the group have more electrons, and so they have more electron shells in their atoms. The outer electrons are further from the nucleus, and so the force between the negatively charged electron and the positively charged nucleus is weaker. It is much easier to remove the outer electron from a caesium atom than it is to remove the outer electron from a lithium atom, so caesium is much more reactive than lithium.

Now try this

1. Write a balanced equation for the reaction of sodium with water. Include state symbols. The symbol for sodium is Na, sodium hydroxide is NaOH, water is H_2O and hydrogen gas is H_2.

(3 marks)

2. Rubidium is an alkali metal below potassium in group 1 of the periodic table. Explain how reactive this metal is compared with potassium.

(3 marks)

Halogens

Group 7

The halogens are the elements in group 7 of the periodic table. They have similar reactions to each other because if they can gain one electron they can each complete their outer shell. They all react with metals to form compounds called halides. Fluorine is the most reactive halogen, and the reactivity decreases as you go down the group.

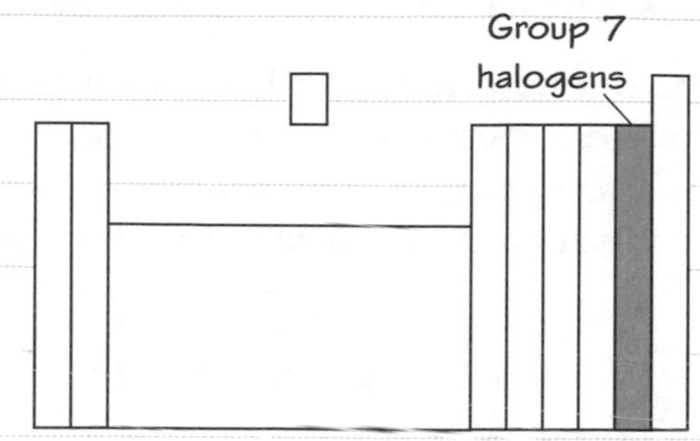

Properties of halogens

Element	Symbol	State at room temperature	Colour
fluorine	F	gas	pale yellow
chlorine	Cl	gas	yellow–green
bromine	Br	liquid	red–brown
iodine	I	solid	grey

You need to learn the states and colours of the halogens.
The symbols in the table are for atoms of the halogens. The halogens form molecules with two atoms in them, so the formulae for the halogens at room temperature are $F_2(g)$, $Cl_2(g)$, $Br_2(l)$ and $I_2(s)$.

Worked example

All the halogens react with metals to form compounds called halides. Write word equations and balanced equations to show the reactions between:

(a) potassium and chlorine **(b)** calcium and bromine.

(a) potassium + chlorine → potassium chloride

$2K(s) + Cl_2(g) \rightarrow 2KCl(s)$

(b) calcium + bromine → calcium bromide

$Ca(s) + Br_2(l) \rightarrow CaBr_2(s)$

You may be asked to write balanced equations for the reactions of any of the halogens with metals. There is more about balancing equations on page 103.

Now try this

1. Chlorine and bromine will both react with sodium. Explain which reaction will happen the fastest. **(2 marks)**

2. Write a word equation and a balanced equation to show the reaction between sodium (Na) and bromine. The formula of sodium bromide is NaBr. **(3 marks)**

More halogen reactions

Reactions with hydrogen

Halogens react with hydrogen to form hydrogen halides. Hydrogen halides dissolve in water to form acidic solutions. For example, when hydrogen chloride is dissolved in water it forms hydrochloric acid.

hydrogen + chlorine → hydrogen chloride

$H_2(g) + Cl_2(g) \rightarrow 2HCl(g)$

All halogens need one electron to complete their outer shell. The formulae for the other hydrogen halides are similar: HF, HBr, HI.

Displacement reactions

Some halogens will displace another halogen from a solution. The table shows how different combinations of halogens and halides react.

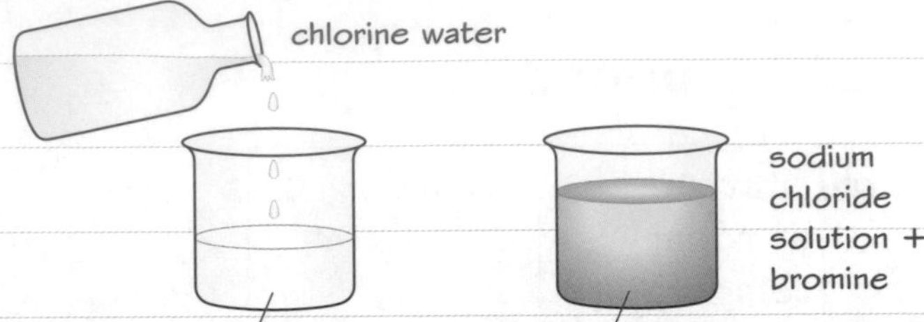

$2NaBr(aq) + Cl_2(aq) \rightarrow 2NaCl(aq) + Br_2(aq)$

halogen	halide ions: fluoride	chloride	bromide	iodide
fluorine		reaction	reaction	reaction
chlorine	no reaction		reaction	reaction
bromine	no reaction	no reaction		reaction
iodine	no reaction	no reaction	no reaction	

Worked example

The table above shows the results of mixing different halogens with solutions containing different halide ions. Explain these results using ideas about reactivity.

Remember that the most reactive halogen is at the *top* of the group. This is the opposite way round to the alkali metals, where the most reactive elements are at the *bottom*.

Fluorine is the most reactive halogen, and the reactivity decreases as you go down the group. Because fluorine is more reactive, it will displace any of the other halogens from their compounds.

Chlorine is less reactive than fluorine, but more reactive than bromine or iodine. If chlorine is mixed with a fluoride solution it will not react, but chlorine will displace bromide or iodide ions from their solutions.

Now try this

1. **(a)** Explain which one of these mixtures will produce a reaction:
potassium bromide + chlorine
potassium chloride + bromine **(3 marks)**

(b) Write a balanced equation for the reaction in part **(a)**. **(2 marks)**

target B-A

2. **(a)** Write a balanced equation to show the reaction between hydrogen and bromine. Include state symbols. **(4 marks)**

(b) Describe how to make the product of this reaction into hydrobromic acid. **(1 mark)**

Noble gases

The noble gases are the elements in group 0 of the periodic table. They all have a full outer electron shell. This is the most stable arrangement of electrons, which other elements only achieve by making ions or sharing electrons in covalent bonds. This is why the noble gases are inert, compared with other elements. This means it is very difficult to make them react.

Uses	Gas	Reason for use
welding	argon or helium	The gases are inert so they do not react with the hot metal, and they stop oxygen coming into contact with it.
inside filament light bulbs	xenon or argon	The gases are inert so they do not react with the hot filament.
balloons and airships	helium	Helium has a low density and makes the balloons float.
fire-extinguishing systems	argon	Argon is non-flammable, so it can be used to put out fires.

Discovery

Lord Rayleigh noticed that the density of nitrogen made in a reaction was lower than the density of nitrogen obtained from air.

Sir William Ramsey hypothesised that the air might also contain a denser gas that was mixed with the nitrogen.

Rayleigh and Ramsey carried out some careful experiments and discovered a new gas – argon.

During the experiments Ramsey discovered helium. Later, he also discovered neon, krypton and xenon.

Worked example

The table shows the densities and boiling points of some of the noble gases. Describe the trends in these properties and use them to predict the missing numbers.

Element	Density (kg/m^3)	Boiling point (°C)
helium (He)	0.15	−269
neon (Ne)	1.20	
argon (Ar)		−186
krypton (Kr)	2.15	−152

The densities increase as you go down the group. The boiling points get higher as you go down the group.

The density of argon could be about 1.7 or 1.8 kg/m^3.

The boiling point of neon could be about −230°C.

These numbers are about halfway between the numbers above and below in the table.

You need to be able to answer questions like this one that ask you to identify a trend, and use the trend to predict missing values.

Now try this

1. (a) Xenon is the next noble gas below krypton in Group 0. Suggest what its density and boiling point could be. **(2 marks)**
   (b) Explain how you worked out the density in part **(a)**. **(2 marks)**
2. Explain which properties are important for the following uses:
   (a) when helium is used in airships **(2 marks)**
   (b) when argon is used in welding. **(2 marks)**

Chemistry extended writing 2

Worked example

Sand is mostly silicon dioxide, with silicon and oxygen atoms joined in a giant molecular covalent structure. Sodium chloride is an ionic compound with the ions held in a lattice structure. Compare the bonding and properties of silicon dioxide and sodium chloride. **(6 marks)**

Sample answer 1

Sand contains covalent bonds and the atoms are joined up in a giant structure. It has high melting and boiling points and does not conduct electricity.

Sodium chloride contains ionic bonds holding the ions together. It has high melting and boiling points and conducts electricity when it is heated into a liquid or dissolved in water.

This is a good answer. It correctly describes the properties of both substances, but does not explain the bonding very well, nor does it explain how the bonding leads to the properties.

Sample answer 2

In sand, silicon and oxygen atoms are joined by covalent bonds, which form when electrons are shared between different atoms. In sodium chloride, each sodium atom forms a positively charged ion by losing an electron, and each chloride atom forms a negatively charged ion by gaining electrons. The attraction of the opposite charges forms ionic bonds that hold the atoms together.

Both substances form giant structures, with billions of atoms (or ions) held together in a lattice structure. Both types of bond are very strong. This means that it takes a lot of energy to break the bonds, so both substances have high melting and boiling points. Silicon dioxide does not dissolve in water, but sodium chloride does.

Silicon dioxide does not conduct electricity, because it does not contain any charged particles that can move around freely. Sodium chloride conducts electricity when it is molten or in solution, as the charged ions can move around freely.

This is an excellent answer. It describes the bonding and properties of both substances, and also points out the similarities and differences between them.

Now try this

1. Explain why different techniques are used to separate oil from a mixture of oil and water, and to separate oxygen from air. **(6 marks)**

Chemistry extended writing 3

Worked example

Different car manufacturers use different paints on their vehicles. Explain how chromatography can be used to identify the make of a car that left a flake of paint behind after a hit-and-run accident.

(6 marks)

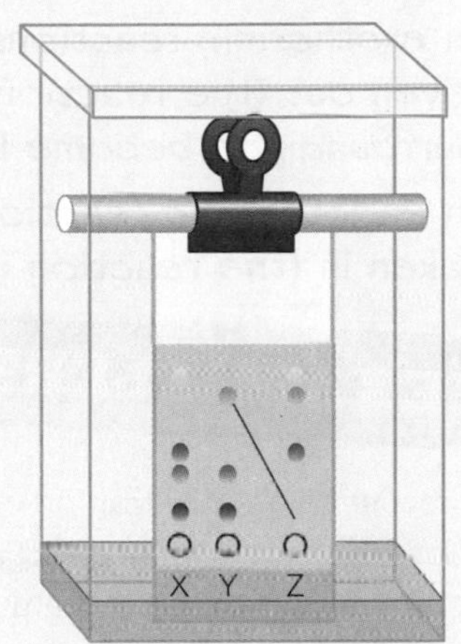

Sample answer 1

Chromatography separates different paints, so put the two paints in water and see how many coloured dots each one makes on the paper. If the two paints match, you know which car it was.

This is a very poor answer. It does not clearly describe how chromatography is carried out, and although it mentions comparing two paints, it does not say what the second paint is.

Sample answer 2

The paint from the accident needs to be compared to paints from different car manufacturers. Paints from different manufacturers will have different mixtures of coloured compounds in them. Dots of each paint are put on the bottom of a piece of paper, then the bottom of the paper is put in a liquid that will dissolve the paint. The colours separate and move up the page. You can tell which paints match because the pattern of dots will be the same.

This is a good answer. It describes the basic method of chromatography, but it could include more detail (such as the paint spots being put *near* the bottom of the paper, not at the bottom), and it could use better terminology (such as 'solvent' instead of 'liquid'). The answer could also be improved by explaining how R_f values are worked out and why they are useful.

Now try this

1. Compare the bonding and properties of carbon dioxide and diamond. **(6 marks)**

Temperature changes

Temperature changes

Most reactions involve a temperature change.

- In exothermic reactions, heat energy is given out (the reaction mixture or the surroundings become hot).
- In endothermic reactions, heat energy is taken in (the reaction mixture gets cold).

EXAM ALERT!

A recent exam question on energy changes was only answered fully by about three out of ten students. Remember: if the temperature increases, the reaction is exothermic. If the temperature decreases, the reaction is endothermic.

Students have struggled with this topic in recent exams - **be prepared!**

Measuring temperature changes

You can use this apparatus to investigate temperature changes in reactions.

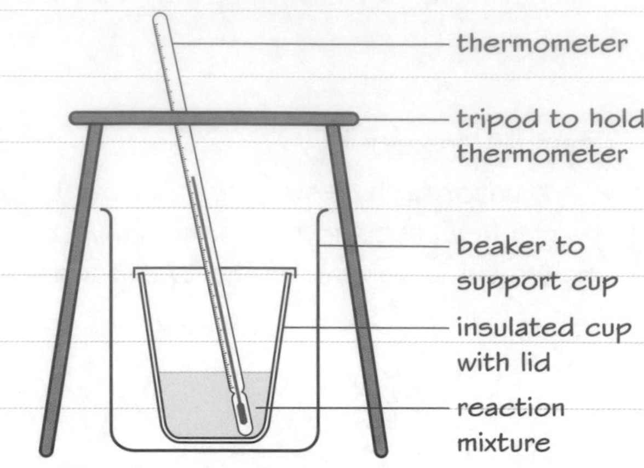

Bonds and energy

When a chemical reaction happens, the bonds that hold the atoms together in the molecules of the reactants are broken. The atoms then come together in new arrangements to form the products.

- Breaking bonds is endothermic (energy is needed).
- Making bonds is exothermic (energy is released).

exothermic reaction

energy

reactants

products

Overall energy is released to surroundings and this makes the reaction exothermic.

Energy is released to the surroundings because more heat energy is released making bonds in the products than is needed to break bonds in the reactants.

endothermic reaction

products

reactants

Overall energy is taken into the reaction and this makes the reaction endothermic.

Energy is taken in from the surroundings because less heat energy is released making bonds in the products than is needed to break bonds in the reactants.

Worked example

Cold-packs are used to relieve muscle pains. Two different substances are mixed, and the mixture gets cold. Explain what is happening in this reaction.

This is an endothermic reaction. The pack feels cold because it is taking in energy from its surroundings.

This is because the energy needed to break the bonds in the reactants must be more than the energy released when the bonds form in the products.

Remember:
endothermic = energy in (energy enters)
exothermic = energy out (energy exits)

Now try this

target A-B

1. Methane burns in air to form carbon dioxide and water.
   (a) Describe whether this is an exothermic or an endothermic reaction. **(2 marks)**
   (b) Draw a labelled energy diagram to illustrate these energy changes. **(2 marks)**

2. When baking powder is mixed with vinegar the temperature of the mixture drops. Explain why this happens, using ideas about making and breaking bonds. **(4 marks)**

Rates of reaction 1

Reactions can happen at very different rates. It can take months for a piece of iron to go rusty, but it only takes a fraction of a second for a mixture of hydrogen and oxygen to explode.

Reactions

Reactions happen when particles of different substances collide with each other. The particles need to collide with enough energy to make the reaction happen, so not all collisions result in a reaction. The higher the frequency of the collisions, and the higher the energy of the collisions, the faster the rate of reaction.

Changing the rate of a reaction does not change the amount of the products formed, it just changes how fast they are formed.

A *faster* reaction is one that happens in a *shorter* time. Don't write that a reaction happens in a faster time – that is not a rate of reaction.

Changing the rate of a reaction

Temperature: Reactions happen faster when the temperature is higher. Particles move faster at higher temperatures, so they collide more often and they also collide with more energy.

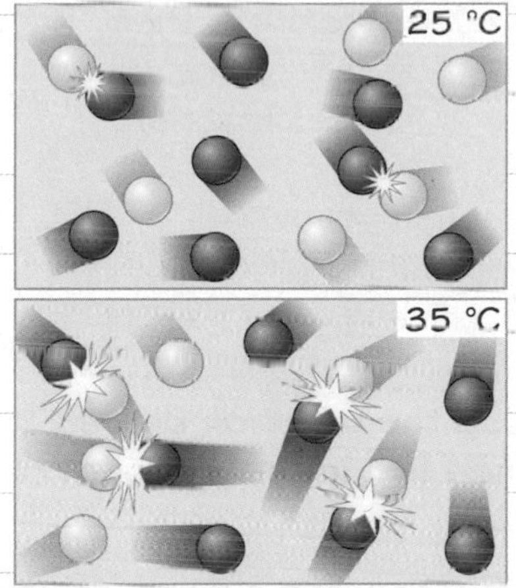

EXAM ALERT!

Less than one in ten students got full marks on a recent exam question about this. Don't forget that higher temperatures mean collisions happen with more energy, as well as there being more collisions.

Students have struggled with exam questions similar to this - **be prepared!**

Surface area: Reactions happen faster when solid reactants are broken up into smaller pieces. Smaller pieces have a bigger surface area than the same mass of larger pieces, so there is more opportunity for collisions between reactants.

Concentration: Reactions happen faster when more concentrated solutions are used. A more concentrated solution has more solute particles in a given volume. The more particles there are, the more likely they are to collide and react.

Catalyst: A catalyst speeds up a reaction without being used up itself.

larger piece

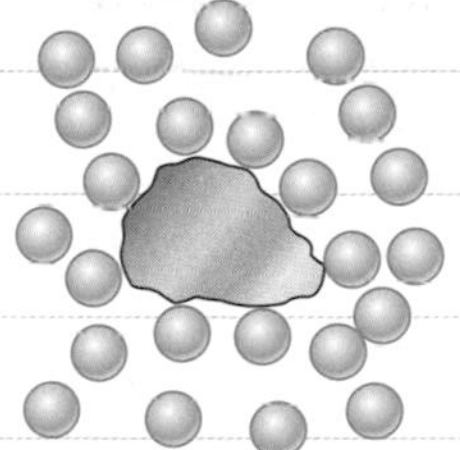

smaller surface area
slower reaction

smaller piece

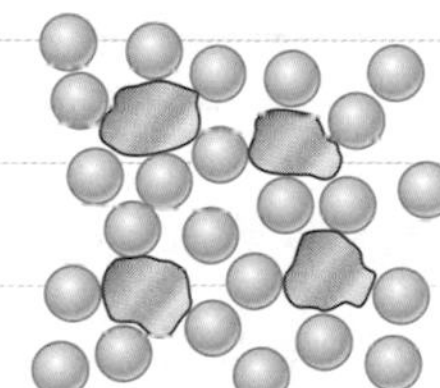

larger surface area
faster reaction

Now try this

target D-C

1. Nickel is used as a catalyst in the manufacture of margarine. A manufacturer uses 1 kg of nickel to manufacture a batch of margarine. Explain how much nickel is left after the reaction. **(2 marks)**

target B-A

2. Explain how each of these changes will change the rate of a reaction:
   (a) using the same mass of a solid reactant, but using larger pieces **(3 marks)**
   (b) carrying out the reaction in colder conditions **(4 marks)**
   (c) using more concentrated alkali in a reaction. **(3 marks)**

Rates of reaction 2

Investigating rates of reaction

Worked example

Two students used this apparatus to investigate the factors that affect the rate of reaction. The magnesium ribbon reacts with the acid to produce hydrogen gas.

Explain how the students could use the apparatus shown here to investigate the effect of temperature on the rate of reaction.

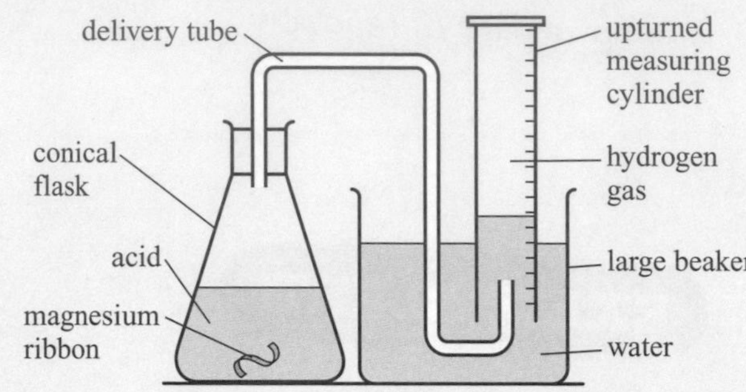

They need to use the same length of magnesium ribbon and the same volume and concentration of acid each time.

They should fill the measuring cylinder with water, then add the magnesium to the acid and put the bung in the conical flask. They should measure the volume of the hydrogen in the measuring cylinder every minute for 5 minutes. They should then repeat the experiment with the acid at different temperatures. They should plot all the results on the same graph to compare them. The graph with the steepest line will be the one with the fastest reaction.

When you are describing how to carry out a practical investigation, remember to say which variables you need to control.

Catalytic converters

Catalytic converters are used in cars to reduce the pollution caused by waste gases from the engine. Carbon monoxide and unburned fuel in the exhaust combine with oxygen to form carbon dioxide and water.

The platinum catalyst in catalytic converters is made into a fine mesh to give it a large surface area. This allows more of the waste gases to come into contact with the catalyst and so helps to speed up the reaction.

Catalytic converters work best at high temperatures.

Now try this

1. Explain why cars fitted with catalytic converters produce more carbon monoxide when they have just started their journey. **(3 marks)**

A sketch graph does not include numbers. It is the shape of the lines on the graph that is important.

2. The reaction described in the worked example above can also be followed by carrying out the reaction in a conical flask standing on a balance. As hydrogen gas is produced by the reaction, the mass of the reactants and products left in the flask decreases.

   Sketch a labelled graph of mass against time showing the results you would expect if a student used powdered magnesium in one reaction and magnesium ribbon in another (keeping all the other factors the same). **(2 marks)**

Relative masses and formulae

Relative formula mass

The masses of atoms are so small that we do not talk about their mass in kilograms. Instead we use relative atomic mass (A_r) (see page 38). The relative formula mass (M_r) is the sum of the relative atomic masses of all the atoms or ions in its formula.

oxygen molecule – formula O_2
relative atomic mass of oxygen = 16
relative formula mass = 2 × 16
= 32

Relative atomic masses and relative formula masses do not have units.

Worked example

Calculate the relative formula mass of calcium nitrate, $Ca(NO_3)_2$.

Relative atomic masses, Ca = 40, O = 16, N = 14

$Ca(NO_3)_2 = 1 \times Ca + 2 \times (1 \times N + 3 \times O)$

$= 1 \times Ca + 2 \times N + 6 \times O$

$M_r = 40 + (2 \times 14) + (6 \times 16)$

$= 164$

You do not need to learn any relative atomic masses. They will always be given to you in an exam question and you will get a copy of the periodic table too.

Empirical formulae and molecular formulae

The empirical formula of a substance is the simplest whole number ratio of atoms or ions of each element in the substance. The empirical formula of sodium chloride is NaCl. This means that for every ion of sodium in an ionic lattice there is one chloride ion. For simple molecular compounds, the molecular formula is the actual number of atoms of each element in the molecule. For example, the molecular formula of water is H_2O.

The empirical formula is not always the same as the molecular formula. For example, ethene molecules (C_2H_4) each have two carbon atoms and four hydrogen atoms. Propene molecules (C_3H_6) each have three carbon atoms and six hydrogen atoms. The empirical formula for both of these compounds is CH_2.

Finding the empirical formula

If you know the mass of the magnesium heated in a crucible and the mass of the magnesium oxide formed then you can work out the mass of oxygen that has combined with the magnesium. From that you can work out the empirical formula.

Now try this

1. Calculate the relative formula masses of the following compounds.
   (a) calcium oxide, CaO **(1 mark)**
   (b) ammonia, NH_3 **(2 marks)**
   (c) calcium chloride, $CaCl_2$ **(2 marks)**
   (Relative atomic masses: Ca = 40, Cl = 35.5, H = 1, N = 14, O = 16)

2. The diagram shows a molecule of hydrogen peroxide. Write down:
   (a) the molecular formula **(1 mark)**
   (b) the empirical formula. **(1 mark)**

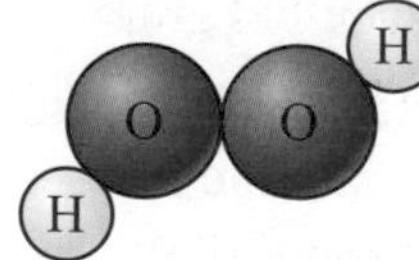

Empirical formulae

Calculating empirical formulae

Worked example

A sample of magnesium chloride contains 5.0 g of magnesium and 14.8 g of chlorine. Calculate the empirical formula. (Relative atomic masses: Mg = 24, Cl = 35.5)

Mg: $\frac{5.0}{24} = 0.208$ Cl: $\frac{14.8}{35.5} = 0.417$

Divide each mass by the relative atomic mass of the element.

Mg: $\frac{0.208}{0.208} = 1$ Cl: $\frac{0.417}{0.208} = 2.005$ (round this to 2)

Divide *both* numbers by the smallest number to find the ratio. Round the answers if necessary.

Empirical formula: $MgCl_2$

Percentage composition

The percentage composition by mass of a compound is a way of working out how much of an element is present in a compound.

$$\text{percentage by mass of an element in a compound} = \text{number of atoms of element} \times \frac{\text{relative atomic mass } (A_r)}{\text{relative formula mass } (M_r)} \times 100$$

Worked example

Calculate the percentage by mass of oxygen in copper carbonate ($CuCO_3$).

(Relative atomic masses: Cu = 63.5, C = 12, O = 16)

There are 3 atoms of oxygen in the formula.

A_r for oxygen = 16

M_r for $CuCO_3$ = 63.5 + 12 + (3 × 16)

= 123.5

Percentage of oxygen by mass = $3 \times \frac{16}{123.5} \times 100$

= 38.9%

Check that your answer makes sense – the percentage you calculate should always be less than 100%.

Now try this

1. A 10.78 g sample of sodium oxide contains 8 g of sodium and 2.78 g of oxygen. Calculate the empirical formula. (Relative atomic masses: Na = 23, O = 16) **(3 marks)**
2. Calculate the percentage by mass of carbon in methane (CH_4). (Relative atomic masses: C = 12, H = 1). **(3 marks)**

target C-A

3. Calculate the percentage by mass of carbon in ethanol (C_2H_5OH). (Relative atomic masses: C = 12, H = 1, O = 16) **(3 marks)**

Masses of reactants and products

No atoms are created or destroyed in a reaction, so the total mass of the products must be equal to the total mass of the reactants. If you know the mass of one of the reactants or products in a reaction, you can use the balanced equation to work out the masses of the other substances involved.

Worked example

0.5 g of lithium metal is added to water. Calculate the mass of hydrogen gas produced. (Relative atomic masses: Li = 7, H = 1)

Always start with the balanced equation. You do not need state symbols.

$2Li + 2H_2O \rightarrow 2LiOH + H_2$

Work out the relative masses and multiply by the balancing numbers. You only need to do this for the substances asked for in the question.

2 x 7   1 x (1 + 1)

14   2

14 g of Li → 2 g of H_2

Divide by the number for lithium to find the mass of hydrogen produced for each gram of lithium.

1 g → $\frac{2}{14}$ g = 0.143 g

Multiply by the mass of lithium given in the question.

0.5 g → 0.143 g x 0.5

= 0.072

0.072 g of hydrogen is produced.

EXAM ALERT!

In recent exam questions, over half of students got no marks at all on a question like this. Always show your working, so even if you make a mistake you may still get some marks. Ask your teacher for help if you don't understand how to do these calculations.

Students have struggled with this topic in recent exams - **be prepared!** ResultsPlus

Now try this

1. Methane (CH_4) burns in oxygen to form carbon dioxide and water.
   $CH_4 + 2O_2 \rightarrow CO_2 + 2H_2O$
   Calculate the mass of oxygen used when 10 g of methane burns.
   (Relative atomic masses: C = 12, H = 1, O = 16) **(4 marks)**

target C-A*

2. Potassium reacts with chlorine to form potassium chloride (KCl).
   Calculate the mass of potassium needed to produce 20 g of potassium chloride. (Relative atomic masses: K = 39, Cl = 35.5) **(4 marks)**

Yields

The yield of a reaction is the amount of useful product you get at the end.

Actual and theoretical yields

You can calculate the mass of product you expect to get from a reaction using the balanced equation and the relative masses of the compounds. This is the theoretical yield of a reaction.

Reactions do not always produce the amount of product predicted from the equation. The amount actually produced is called the actual yield.

Remember, the actual yield is sometimes the same as the theoretical yield, but usually it is less. The actual yield is never *more* than the theoretical yield.

The formula is

percentage yield $= \frac{\text{actual yield}}{\text{theoretical yield}} \times 100\%$

You may be given yields in kilograms instead of tonnes. It doesn't matter what the units are, as long as the units for the theoretical and actual yields are the same. Your answer will always be a percentage.

Worked example

Some limestone is heated to produce calcium oxide. The theoretical yield is 50 tonnes of calcium oxide. The actual yield is 30 tonnes. What is the percentage yield?

$$\text{percentage yield} = \frac{30 \text{ tonnes}}{50 \text{ tonnes}} \times 100\%$$

$$= 60\%$$

Why don't we always get full yield?

- A reaction does not always finish – there may be some reactants left at the end.
- Some of the reactants or products are lost during the process – such as liquids being spilled or left behind in a container.
- There may be unwanted reactions taking place – some reactants may react in a different way to make a different product.

Now try this

1. Explain the difference between theoretical yield and actual yield. **(2 marks)**
2. Iron is extracted from iron oxide. The theoretical yield of the reaction is 600 tonnes. The actual yield is 240 tonnes. Calculate the percentage yield. **(2 marks)**

Waste and profit

Most reactions have more than one product. Most reactions in the chemical industry produce substances other than the substance being manufactured. These are by-products.

Worked example

Explain three ways in which chemists in industry try to find the most economically favourable reactions.

They try to find:

- reactions with high yields, so that a lot of product is made in the reaction
- reactions where all the products are commercially useful, as they will make more money if they can sell all the products and do not have to dispose of waste products
- reactions that occur fairly quickly, so that useful products can be made quickly.

Disposing of waste products

House prices could drop if there is a new chemical plant built, or if there are unpleasant smells.

Some waste products may accidentally escape and cause pollution.

People do not like living near landfill sites or incinerators.

We are running out of landfill sites.

Many waste products that are not harmful still need to be disposed of. They may have to be transported to a landfill site or incinerator. Most commercial companies have to pay to use these facilities.

Some waste products are harmful, and treating them to make them safe costs money.

The smoke from incinerators is not as harmful as the substances that have been burned, but incinerating waste can still cause some air pollution.

Lorries taking waste to landfill sites or incinerators cause dust and disturbance.

Not all waste products are harmful. Some reactions produce water as a waste product, and this can be disposed of easily as long as it does not contain other substances.

Now try this

1. When iron is extracted from iron ore commercially, a waste product called slag is produced. Slag can be used to make cement. Explain how this might help the iron industry to make a profit. **(3 marks)**

Chemistry extended writing 4

Worked example

Sodium thiosulfate solution reacts with hydrochloric acid to form a precipitate of sulfur. As the sulfur forms, the mixture gradually becomes more difficult to see through. Changing the concentration of the hydrochloric acid changes the time it takes for the mixture to become too cloudy to see through.

Explain how you could use this apparatus to investigate the effect of the concentration of acid on the rate of the reaction. **(6 marks)**

Sample answer 1

Use different concentrations of acid and see how long the reaction takes each time.

This is a basic answer. It does not contain enough detail about how to judge when the reaction is finished, or how to make sure the investigation is a fair test.

Sample answer 2

Time how long the reaction takes by measuring how long it takes for the cross to disappear. Then wash out the flask and put the same volume and concentration of sodium thiosulfate into it. Use a different concentration of acid but use the same volume, and time the reaction again. Do this with three more different concentrations of acid. The temperature should be the same each time.

This is a good answer. There is a good list of the variables that must be kept the same to make the test fair. However, judging when the cross disappears may not be easy to do accurately, so this answer could be improved by explaining why repeat measurements should be taken and how the results would be analysed.

Now try this

1. When a chemical reaction takes place there is often a temperature change. Explain how you can find out whether a neutralisation reaction is an endothermic or exothermic change, and explain why the temperature changes. Your answer should include a discussion of bond energies. **(6 marks)**

Chemistry extended writing 5

Worked example

Explain the differences between an empirical formula, a molecular formula and the percentage composition by mass of a compound, and how to work them out. **(6 marks)**

Sample answer 1

The empirical formula is the simplest ratio of the different atoms or ions in a compound. You can work this out by finding the mass of different elements in a compound. For example, if you heat magnesium in air, magnesium oxide forms. You can find the mass of magnesium at the beginning and the mass of magnesium oxide at the end to work out the mass of magnesium and oxygen. You then need to divide the reacting masses by the relative atomic mass of each element, and then find the ratio of the two numbers.

This is a good answer. The explanation of an empirical formula and how to find it is excellent, but an answer needs to answer *all* of the question.

Sample answer 2

A second student described empirical formulae in a similar way, but then added the following:

The molecular formula is the actual number of atoms of each element in a molecule of the substance. For some substances this is the same as the empirical formula, but not always. For example, the empirical formula for all alkenes is CH_2, but ethene has the molecular formula C_2H_4.

The percentage composition by mass is worked out from the empirical or molecular formula. You need to find the relative formula mass of the compound, and the relative atomic masses of the different elements, and then find the percentage.

This is an excellent answer. There is enough here to score very well. The only thing that could be added would be some more details of calculating the percentage composition by mass.

Now try this

1. Explain three different factors that can affect the rate of a reaction. **(6 marks)**

Static electricity

Atoms

Atoms have a nucleus containing protons and neutrons. Electrons move around the nucleus of an atom. An atom has the same number of protons and electrons, so the + and – charges balance and the atom has no overall charge.

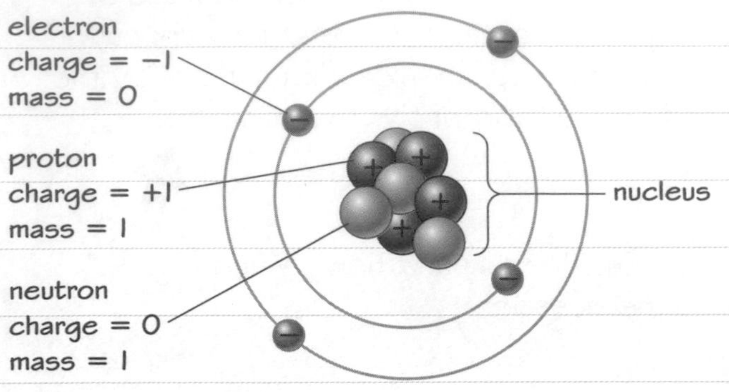

Electrostatic charges

Insulating materials can be given an electrostatic charge by rubbing two materials together. Electrons are transferred from one material to the other. The material that has gained electrons has a negative charge. The material that has lost electrons has a positive charge equal in size to the negative charge.

EXAM ALERT!

In a recent exam question on static electricity about half of students did not achieve any marks. Remember it is always the **electrons** that move. If a material has a positive charge it is because it has *lost* some electrons and so now it has more protons than electrons.

Students have struggled with this topic in recent exams - **be prepared!** ResultsPlus

Charging by induction

A charged object (such as a plastic comb) can attract uncharged objects (such as small pieces of paper). This happens because the comb induces a charge in the pieces of paper.

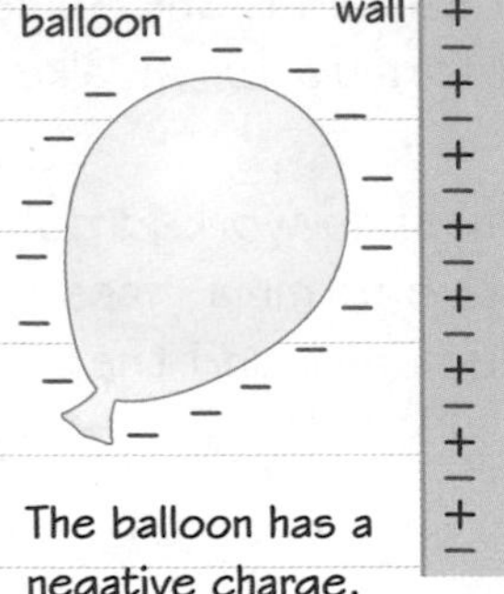

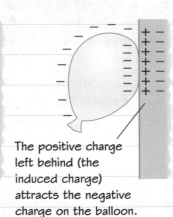

The electrons in the wall are repelled and move away.

The positive charge left behind (the induced charge) attracts the negative charge on the balloon.

Now try this

target C-B

1. You rub a comb with a cloth. The comb gets a positive charge.
   (a) Explain how particles have been transferred to give the comb a positive charge. **(4 marks)**
   (b) Explain what will happen if you hold the comb close to a positively charged rod hanging on a thread. **(2 marks)**

target B-A*

2. You can use a charged comb to pick up small pieces of paper. Explain how this happens, using ideas about charges. **(4 marks)**

Uses and dangers

Shocks and earthing

You can sometimes build up a charge of static electricity as you move around. If you then touch a door or other object you may feel a shock. This happens as electrons move to 'cancel out' the charge on you. This is called earthing.

If you have a negative charge...
...electrons will flow from you to earth.

Lightning happens when a charge of static electricity builds up in clouds. When the charge is big enough, charged particles can flow through the air. The energy released by this causes light and sound. Lightning can kill living things and damage buildings.

Using static electricity

Static electricity is useful in paint spraying and insecticide sprayers.

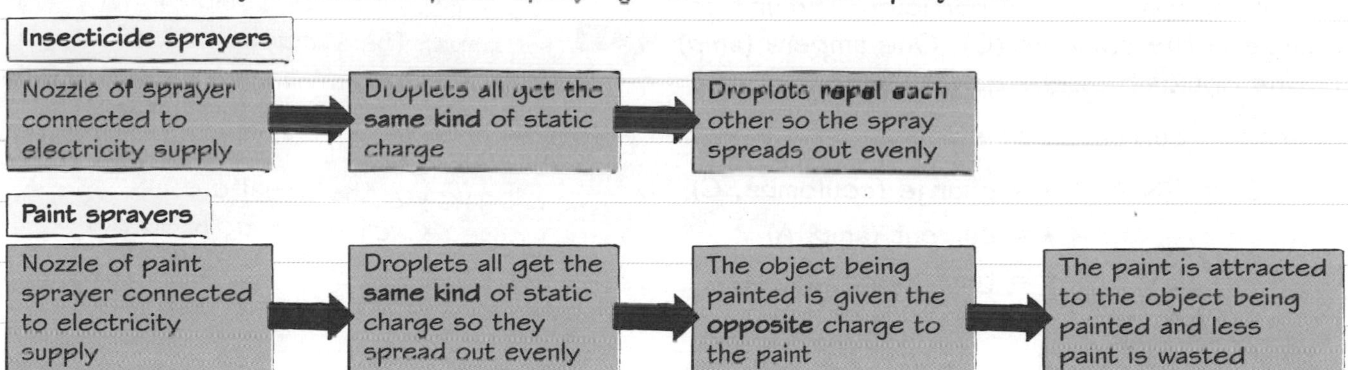

Nozzle of sprayer connected to electricity supply
Droplets all get the same kind of static charge
Droplets repel each other so the spray spreads out evenly

Nozzle of paint sprayer connected to electricity supply
Droplets all get the same kind of static charge so they spread out evenly

Dangers of static electricity

Worked example

Explain how static electricity can be dangerous when aeroplanes are refuelling, and how this problem is solved.

A similar problem can occur when tankers deliver fuel to filling stations. To stop explosions the hose used to fill the tanks is made of a conducting material. Any charge can be earthed without causing sparks.

Static electricity can build up on an aeroplane as it flies. When it is being refuelled, the static charge may cause a spark when the nozzle of a fuel tanker touches the aeroplane. This could cause an explosion if it ignited fuel vapour.

A conducting wire called a bonding line is used to earth any static charge on the aeroplane before refuelling starts. Electrons can flow along the wire to earth to neutralise the static charge on the aeroplane.

Now try this

target D-B

1. You walk across a carpet and then touch a metal doorknob. Explain why you may get a shock. (You can explain either how you would get a positive or negative charge building up.) **(4 marks)**

target D-C

2. Explain how static electricity is used in paint spraying. **(4 marks)**

Electric currents

Direct current

An electric current in a wire is a flow of electrons. The current supplied by cells and batteries is direct current (d.c.). In a direct current the electrons all flow in the same direction.

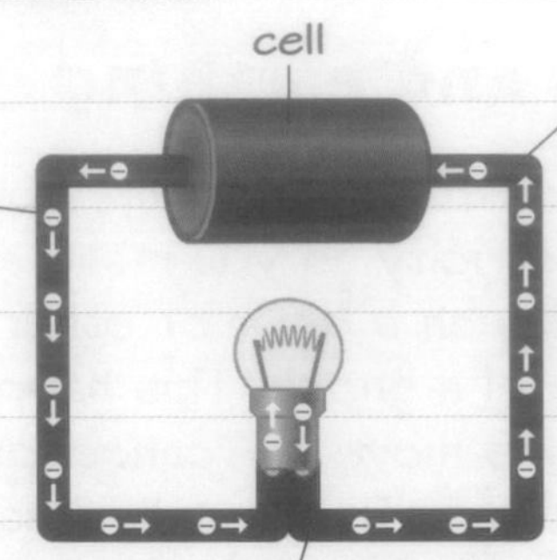

cell
Electrons are pushed out of one end of the cell.
Electrons flow round to the other end of the cell.
There must be a complete circuit for the electrons to flow.

Charge and current

The size of a current is a measure of how much charge flows past a point each second. It is the rate of flow of charged particles. The unit of charge is the coulomb (C). One ampere (amp) is one coulomb of charge per second.

charge = current × time

$Q = I \times t$ Q = charge (coulombs, C)
I = current (amp, A)
t = time (seconds, s)

Don't get the units and quantities confused. The *units* have sensible abbreviations (C for coulombs, A for amps). The symbols for the quantities are not as easy to remember (Q stands for charge, and I for current). The exam paper will give you the equations in words as well as symbols, so don't worry too much if you cannot remember what Q and I stand for.

Worked example

1200 C of charge takes 2 minutes to flow through a bulb. Calculate the current.

Time must be in seconds:

2 minutes = 2 × 60 seconds
= 120 seconds

$$I = \frac{Q}{t} = \frac{1200\,\text{C}}{120\,\text{s}} = 10\,\text{A}$$

You need to remember the correct units for each quantity. The equations will be given to you in the exam, but not the units!

EXAM ALERT!

Always show your working. Even if you get the final answer wrong you may still be able to demonstrate that you understand how to carry out the working.

Students have struggled with exam questions similar to this - **be prepared!** ResultsPlus

Now try this

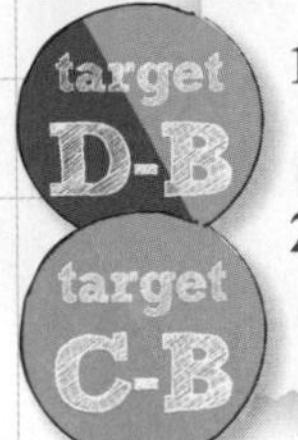

1. 500 C of charge flow through a circuit in 10 seconds. Calculate the size of the current. **(2 marks)**
2. A 5 A current transfers 100 C of charge. Calculate the amount of time the current was flowing. **(3 marks)**

Current and voltage

Measuring current and voltage

The current in an electric circuit is measured using an ammeter. The ammeter is placed in a circuit in series with the other components.

Potential difference is another name for voltage, so you can use either term.

The potential difference (voltage) across a component is measured using a voltmeter. The voltmeter is placed in parallel with the component. The potential difference is the energy transferred by each coulomb of charge that passes through a component. 1 volt is 1 joule per coulomb (1 V = 1 J/C).

You need to learn the symbols for the different circuit components shown here.

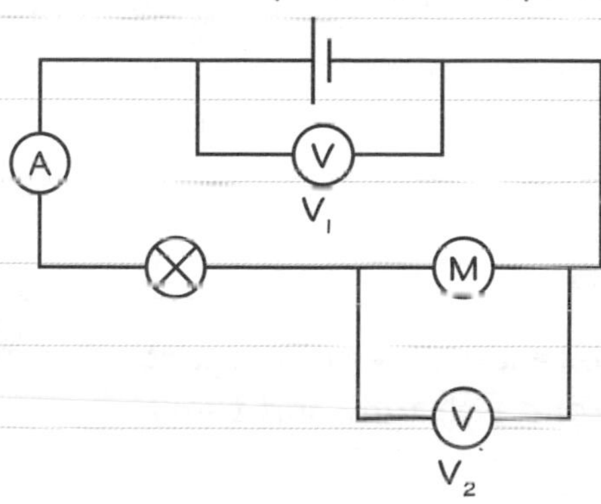

Voltmeter V_1 is measuring the potential difference across the cell. Voltmeter V_2 is measuring the potential difference across the motor.

Worked example

How does the size of the current change if the potential difference (voltage) of the power supply is changed?

If the potential difference is increased, the current increases. If the potential difference is made smaller, the current gets smaller.

Current in parallel circuits

The current in a series circuit is the same everywhere. A parallel circuit has more than one path for the current to flow through. The current splits up when it reaches a junction and comes back together when the wires rejoin.

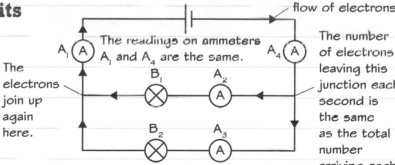

The electrons join up again here.

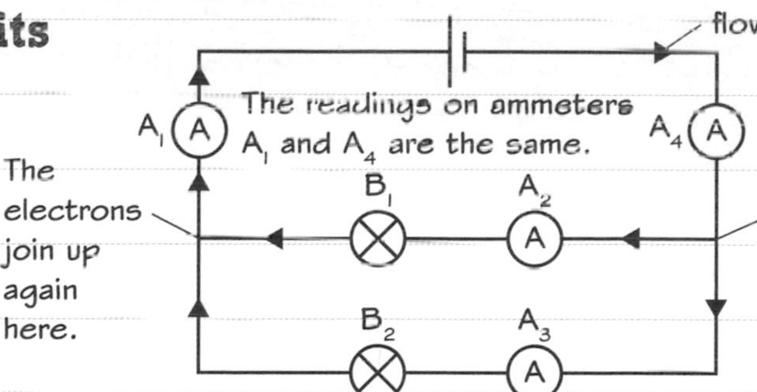

The readings on ammeters A_2 and A_3 add up to the reading shown on A_1 and A_4.

The number of electrons leaving this junction each second is the same as the total number arriving each second.

Now try this

target D-C

1. Draw a series circuit that you could use to measure the current through a bulb and the potential difference across it. **(3 marks)**

2. In the parallel circuit shown above, the reading on ammeter A_1 is 3 A and the reading on ammeter A_2 is 1.2 A. State the reading on:
   (a) ammeter A_4 **(1 mark)**
   (b) ammeter A_3. **(1 mark)**

target C-A

3. The voltage across a component is 5 V. Calculate the energy transferred when 20 C of charge flow through it. **(2 marks)**

Resistance, current and voltage

Resistance

The resistance of a component is a way of measuring how hard it is for electricity to flow through it. The units for resistance are ohms (Ω).

The resistance of a whole circuit depends on the resistances of the different components in the circuit. The higher the total resistance, the smaller the current.

resistance UP, current goes DOWN

The resistance of a circuit can be changed by putting different resistors into the circuit, or by using a variable resistor. The resistance of a variable resistor can be changed using a slider or knob.

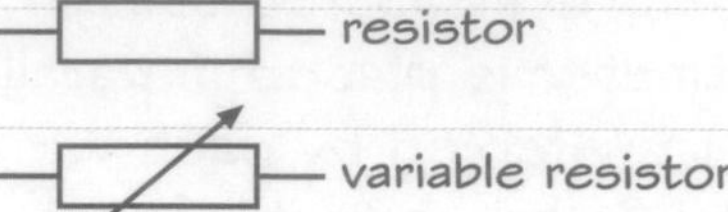

Circuit calculations

You can calculate the resistance of a component by measuring the current and voltage, and then rearranging this formula:

potential difference = current × resistance

$V = I \times R$

V = voltage (or potential difference) (volts, V)

I = current (amps, A)

R = resistance (ohms, Ω)

V
I × R

You don't need to remember this formula, as it will be given to you in the exam. You *do* need to be able to use it, and to remember the correct units for the different quantities.

Worked example

Resistor A has a current of 3 A flowing through it when the voltage across it is 15 V.

What is its resistance?

$$R = \frac{V}{I}$$

$$R = \frac{15\text{ V}}{3\text{ A}} = 5\ \Omega$$

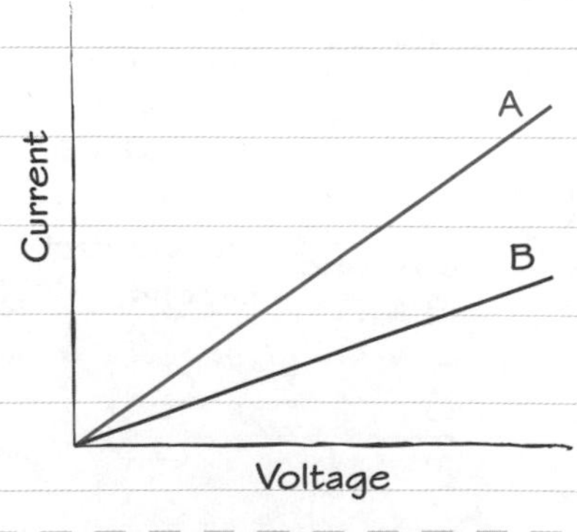

The steeper line shows that resistor A has a lower resistance than resistor B.

A graph of current against voltage for a resistor is a straight line if the temperature of the resistor does not change.

In the example above, resistor A had a current of 3 A when the voltage was 15 V. Resistor B has a higher resistance than Resistor A, so the current will be lower for the same voltage.

Now try this

1. There is a voltage of 30 V across a 15 Ω resistor. Calculate the current. **(3 marks)**
2. A 6 V battery provides a current of 2 A in a circuit. Calculate the resistance of the circuit. **(3 marks)**

Changing resistances

Some electrical components change their resistance depending on the potential difference or the conditions surrounding them.

Filament lamps

Filament lamps get hotter as the voltage across them increases. This increases their resistance. The higher the temperature, the higher the resistance.

voltage UP, resistance UP

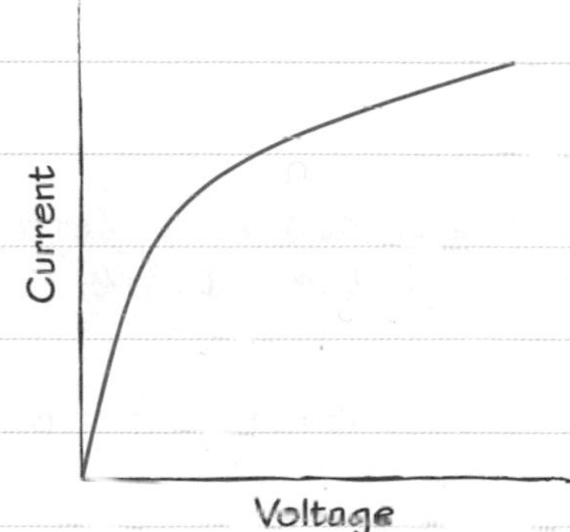

The line gets shallower at higher voltages, showing that the resistance is increasing.

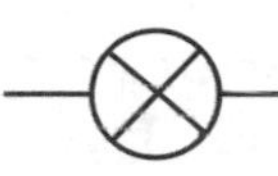

Diodes

When the current flows in one direction diodes behave like fixed resistors. The resistance does not change if the voltage changes. Diodes only conduct electricity in one direction. They do not allow current to flow in the other direction.

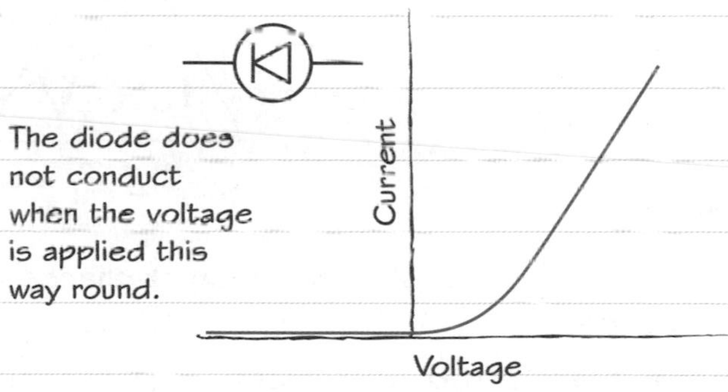

The diode does not conduct when the voltage is applied this way round.
Current
Voltage

Light-dependent resistors

The resistance of a light-dependent resistor (LDR) is large in the dark. The resistance gets less if light shines on it. The brighter the light, the lower the resistance.

brightness UP, resistance DOWN

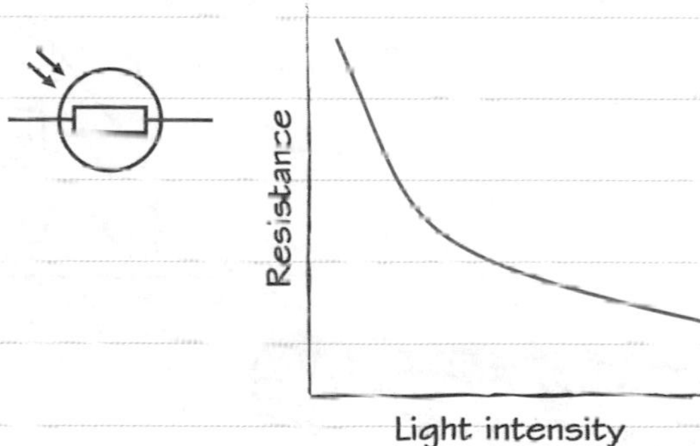

Thermistors

The resistance of a thermistor depends on its temperature. The higher the temperature, the lower the resistance.

temperature UP, resistance DOWN

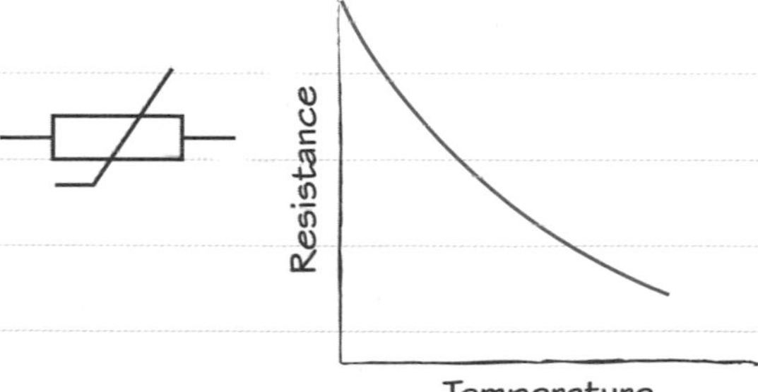

Now try this

1. A current is flowing in a circuit with a fixed-voltage power supply. Explain how the current would change in the following circumstances:
   - **(a)** brighter light shines on an LDR in the circuit **(2 marks)**
   - **(b)** a thermistor in the circuit is put into iced water **(2 marks)**
   - **(c)** the connections to a diode in the circuit are swapped over. **(2 marks)**

Transferring energy

Energy is transferred to a resistor when a current flows through it. This energy transfer heats the resistor. The heating effect is useful in electric fires or electric cookers. Some heating is not useful, such as when energy is wasted as heat energy in motors or other electric components.

Power

The power of an appliance is the energy transferred per second. The unit for power is the watt (W).
1 watt = 1 joule per second.

Don't get watts and joules mixed up. Remember: joules are for energy, watts are for power (energy per second).

electrical power (watt, W) = current (amp, A) × potential difference (volt, V)

$P = I \times V$

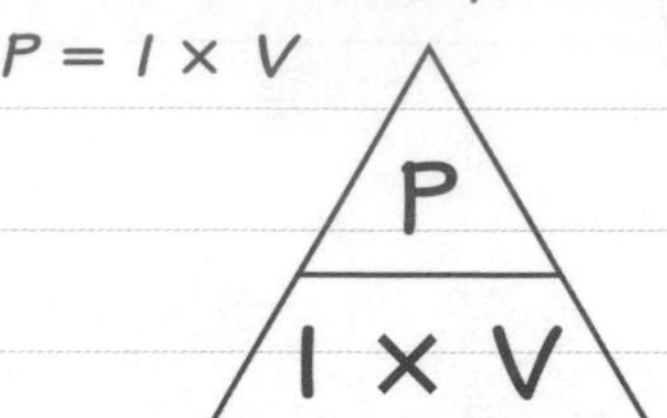

Energy

The total energy transferred by an appliance depends on its power and how long it is switched on.

energy transferred (joule, J) = current (amp, A) × potential difference (volt, V) × time (second, s)

$E = I \times V \times t$

E
I × V × t

Worked example

A travel kettle transfers 10 800 J of energy in one minute, using electricity from a 12 V car battery. Calculate the current.

1 minute = 60 seconds

$$\text{current} = \frac{E}{V \times t} = \frac{10\,800\text{ J}}{12\text{ V} \times 60\text{ s}} = 15\text{ A}$$

Be careful with your units. If you are given a time in minutes, you must convert it to seconds by multiplying by 60.

Now try this

1. A heater uses the mains supply of 230 V. The current through it is 9 A. Calculate the energy it transfers in 10 minutes. State the units. **(4 marks)**
2. A 3 W motor is connected to a 6 V cell. Calculate the current. **(3 marks)**

Physics extended writing 1

To answer an extended writing question successfully you need to:

- ✓ use your scientific knowledge to answer the question
- ✓ organise your answer so that it is logical and well ordered
- ✓ use full sentences in your writing and make sure that your spelling, punctuation and grammar are correct.

Worked example

A student rubs a polythene rod with a cloth and then holds it close to some small pieces of tissue paper. The pieces of paper are attracted to the rod. She repeats the experiment using a metal rod, but this does not attract the tissue paper.

Explain why this happens. **(6 marks)**

Sample answer 1

Rubbing the polythene rod charges it up, and it induces a charge in the pieces of tissue paper. You cannot give metals a charge.

This is a basic answer. It contains some correct information, but there is not enough detail.

Sample answer 2

When the polythene rod is rubbed, electrons are transferred from the cloth to the rod so that it becomes negatively charged. The cloth will have a positive charge.

When the charged rod is held close to the pieces of paper the negative charge pushes electrons in the paper away from the surface. The surface then has a positive charge, which is called an induced charge. The paper is attracted to the negatively charged rod because opposite charges attract.

When you rub a metal rod, electrons could be transferred but the charge will spread out. This is because the metal is a conductor. There is not a big enough static charge on the end of the rod to induce a charge in the tissue paper.

This is an excellent answer. It includes all the details needed and uses correct scientific words such as 'induced charge'.

Now try this

1. A 5 Ω resistor is put into a circuit with a voltmeter, an ammeter and a power supply. A filament lamp with the same resistance is put into a similar circuit. Explain the way that the meter readings in the two circuits would change as the voltage of the power supply is increased. **(6 marks)**

You could provide a sketch graph to help you to explain your answer. A sketch graph has labelled axes and shows the shape of the curve. It does not need to have numbers on the axes.

Vectors and velocity

Vectors

Some quantities are vectors. They have a direction as well as a size. Vectors include:

- displacement
- velocity
- force
- acceleration.

Calculating speed

Speed is calculated from a distance and a time.

$$\text{speed (m/s)} = \frac{\text{distance (m)}}{\text{time (s)}}$$

Distance–time graphs

A student is riding their bike. The distance–time graph tells us about their journey.

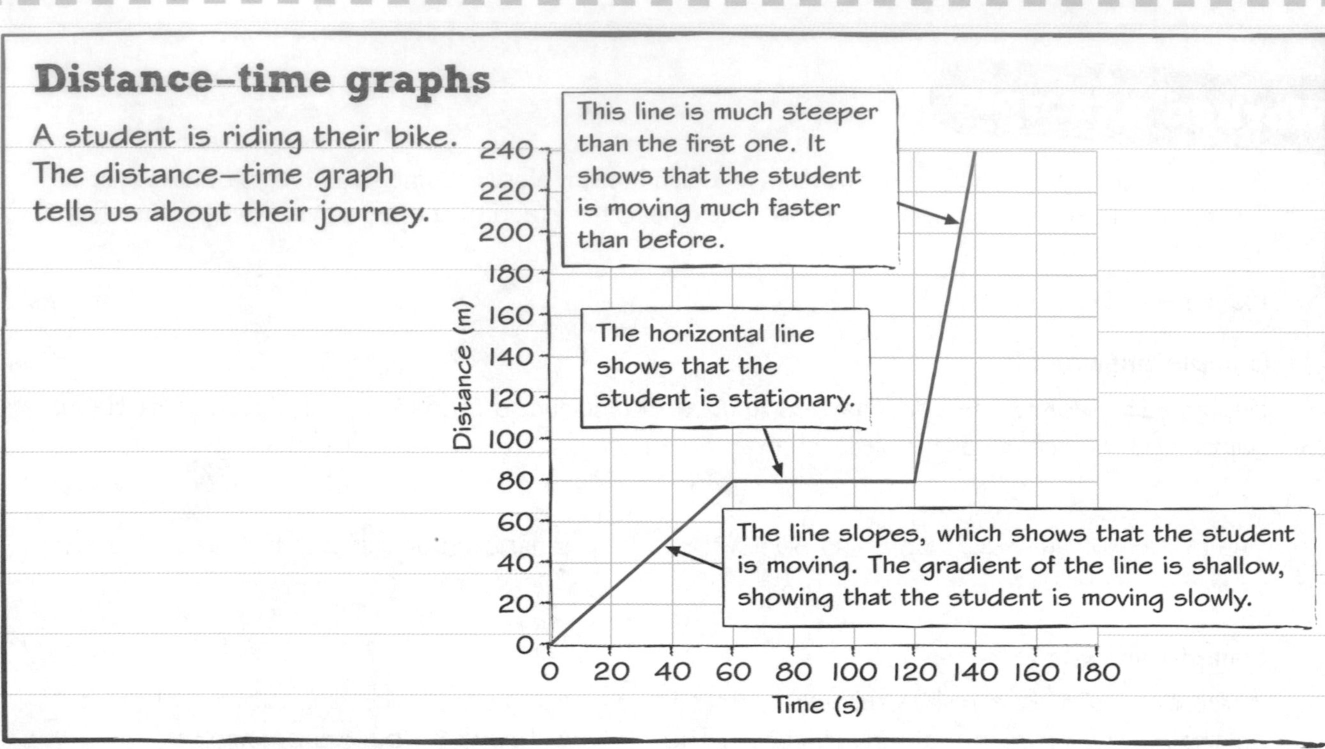
This line is much steeper than the first one. It shows that the student is moving much faster than before.
The horizontal line shows that the student is stationary.
The line slopes, which shows that the student is moving. The gradient of the line is shallow, showing that the student is moving slowly.
Distance (m)
0
20
40
60
80
100
120
140
160
180
200
220
240
Time (s)
0
20
40
60
80
100
120
140
160
180

Worked example

Calculate the student's speed during the last part of their journey, using the gradient of the line on the graph.

Distance = 240 m − 80 m = 160 m

Time = 140 s − 120 s = 20 s

Find the change in distance from the graph, and the change in time.

$$\text{Speed} = \frac{160\text{ m}}{20\text{ s}} = 8\text{ m/s}$$

Use these numbers in the equation for speed.

Now try this

target D-C

1. Calculate the student's speed for the first part of their journey. **(4 marks)**
2. A car is travelling around a roundabout at a constant speed. Explain why the car is said to be accelerating. **(3 marks)**

Velocity and acceleration

Acceleration

Acceleration is a change in velocity.
Acceleration is a vector quantity.

$$\text{Acceleration (m/s}^2) = \frac{\text{change in velocity (m/s)}}{\text{time taken (s)}}$$

$$a = \frac{(v - u)}{t}$$

a is the acceleration
v is the final velocity
u is the initial velocity
t is the time taken

Worked example

A car travelling at 20 m/s slows down to 10 m/s in 4 seconds.
Calculate its acceleration.

$$\text{acceleration} = \frac{(10\text{ m/s} - 20\text{ m/s})}{4\text{ s}} = \frac{-10\text{ m/s}}{4\text{ s}} = -2.5\text{ m/s}^2$$

Slowing down is also an 'acceleration'. An object slowing down will have a negative acceleration. If you remember to always put the final velocity first in the equation, your answer will have the correct sign.

Velocity–time graphs

This velocity–time graph shows how the velocity of a train along a straight track changes with time. The distance travelled can be calculated from the area under the line.

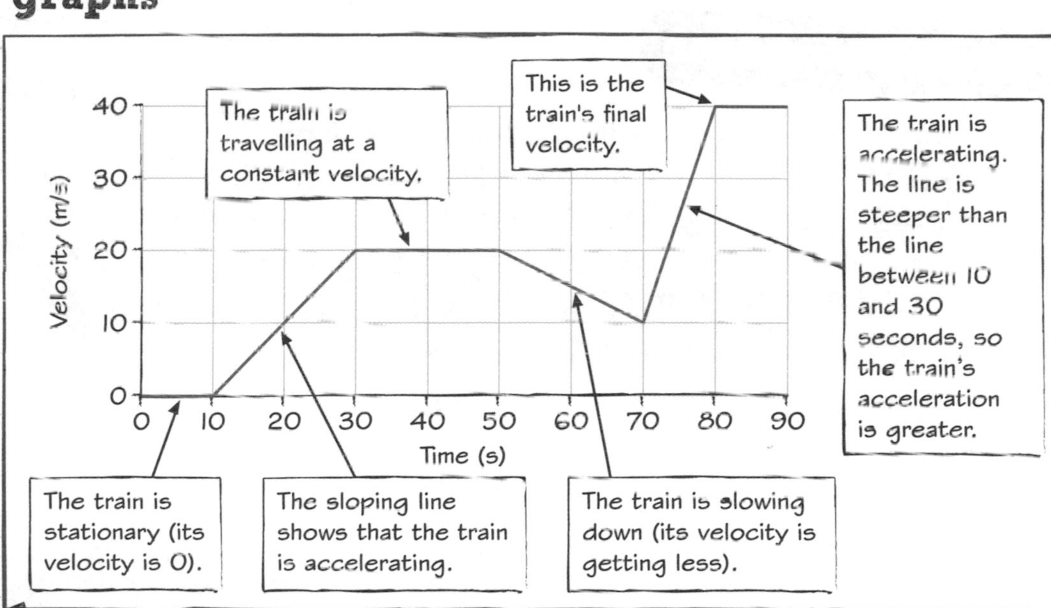
Velocity (m/s)
Time (s)
0
10
20
30
40
50
60
70
80
90
The train is travelling at a constant velocity.
This is the train's final velocity.
The train is accelerating. The line is steeper than the line between 10 and 30 seconds, so the train's acceleration is greater.
The train is stationary (its velocity is 0).
The sloping line shows that the train is accelerating.
The train is slowing down (its velocity is getting less).

Watch out! Check whether you are looking at a distance–time or velocity–time graph. The slope of the line gives speed on a d–t graph and acceleration on a v–t graph.

You need to remember how to work out the area of a triangle:
$\text{area} = \frac{1}{2} \times \text{base} \times \text{height}$

Now try this

target D-C

1. Calculate the acceleration of the train between 50 and 70 seconds. **(3 marks)**

Read the numbers you need off the graph.

target C-A*

2. Calculate the distance the train travels between 70 and 90 seconds. **(4 marks)**

Resultant forces

If there is more than one force on a body, all the forces can be combined into a resultant force.

Free-body diagrams

A force is a vector quantity, because it has a direction as well as a size. A free-body force diagram represents all the forces on a single body. Larger forces are shown using longer arrows.

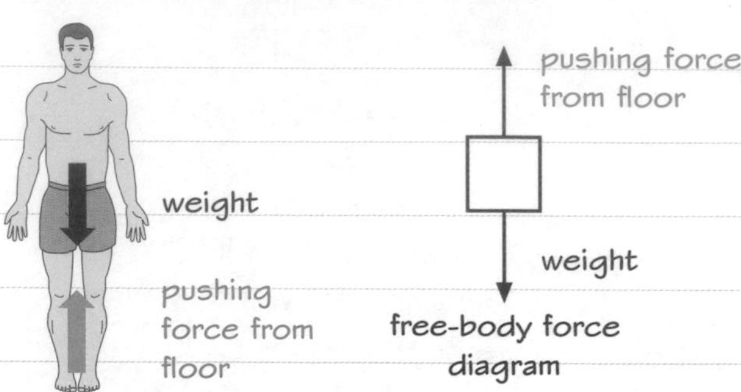

Action and reaction forces

Two touching objects exert forces on each other.

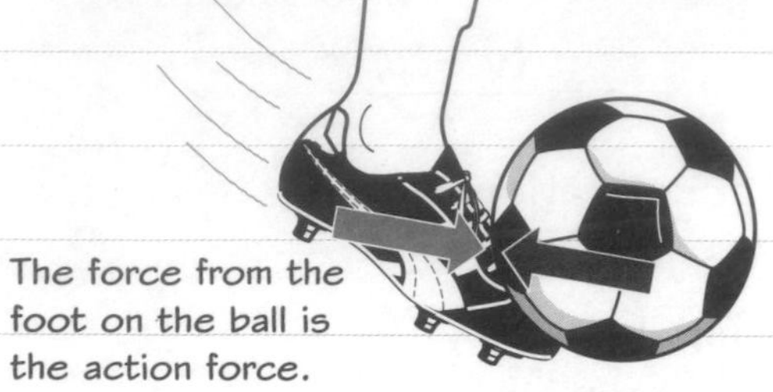

The force from the foot on the ball is the action force.
The ball exerts an equal and opposite reaction force on the boot.

Watch out! Free-body diagrams are for forces on the same body. Action and reaction forces are forces on separate bodies.

Worked example

Explain the effect that each of these forces will have on a car.

(a) 300 N forward force from the engine, 200 N drag.

(a) Resultant force = 300 N − 200 N = 100 N. The car will accelerate in the direction of the resultant force. Its velocity will increase.

A resultant force acting in the opposite direction to the movement of a body will slow it down.

(b) 200 N forward force from the engine, 400 N friction from brakes.

(b) Resultant = 200 N − 400 N = −200 N (200 N acting backwards). The car will accelerate in the direction of the resultant force. This is in the opposite direction to its velocity, so the car will slow down.

(c) 300 N forward force, 300 N drag.

(c) Resultant = 0 N, so the car will continue to move at the same velocity.

Now try this

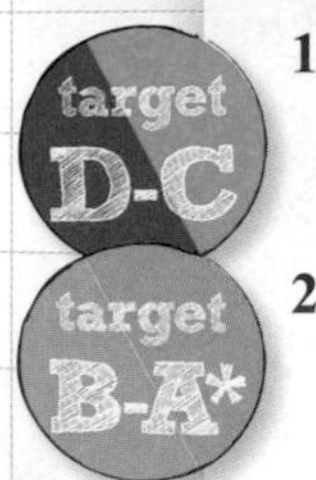

1. A girl is sitting on a chair.

(a) Draw a free-body diagram to represent the forces on her. **(2 marks)**

(b) Describe an action–reaction pair of forces for the girl and the chair. **(2 marks)**

2. An aeroplane is flying with balanced forces on it. It drops a crate of supplies. Explain how this will affect the resultant force on the aeroplane:

(a) in a vertical direction **(2 marks)**

(b) in a horizontal direction. **(2 marks)**

Forces and acceleration

Force and mass

The acceleration produced by a resultant force depends on the size of the force and the mass of the object.

- The greater the force, the greater the acceleration for a constant mass.
- The greater the mass, the smaller the acceleration for a constant force.

Investigating acceleration

You can investigate the effects of mass on acceleration using apparatus like this.

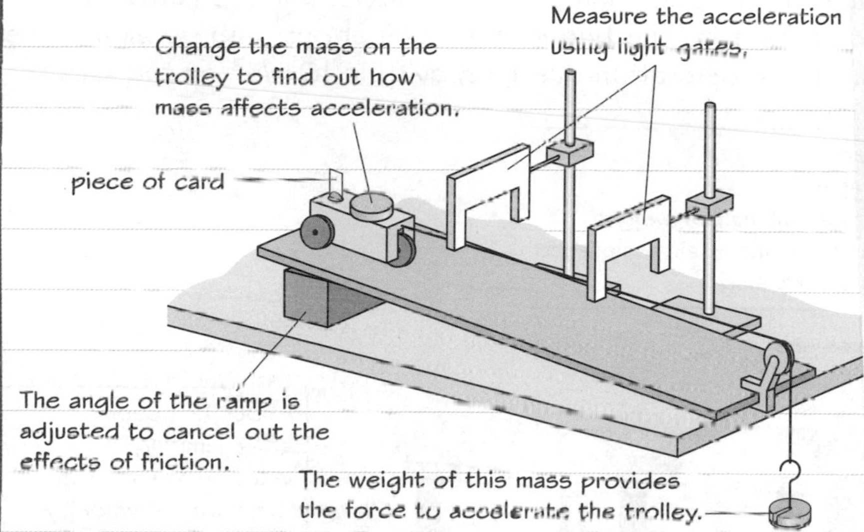

Measure the acceleration using light gates.
Change the mass on the trolley to find out how mass affects acceleration.
piece of card
The angle of the ramp is adjusted to cancel out the effects of friction.
The weight of this mass provides the force to accelerate the trolley.

You can also use this apparatus to investigate the effect of force on acceleration. You can increase the force by increasing the number of masses on the end of the string. The masses on the pulley are accelerating as well as the trolley, and you need to keep the total mass constant to make sure the test is fair. The best way to do this is to start with several masses on the trolley, and transfer these to the pulley one by one to increase the force.

Calculating acceleration

Worked example

A car has a mass of 1200 kg. It accelerates at 2.5 m/s^2. Calculate the force provided by its engine using the formula force (N) = mass (kg) × acceleration (m/s^2).

$F = m \times a$

$F = 1200\text{ kg} \times 2.5\text{ m/s}^2$

$= 3000\text{ N}$

Remember that there will also be forces of friction and air resistance acting, so in real life a car engine would need to produce a larger force than this to achieve 2.5 m/s^2 acceleration.

Now try this

1. A flea undergoes an acceleration of 1.32×10^3 m/s^2 when it jumps. The flea has a mass of 0.45 mg (0.45×10^{-3} kg). Calculate the force exerted by the flea's legs. **(2 marks)**

target C-B

2. A greyhound accelerates at 5 m/s^2 with a force of 150 N. Calculate its mass. **(3 marks)**

Terminal velocity

Mass and weight

Mass:
- is the amount of matter in an object
- is measured in kilograms (kg).

Weight:
- is the force of gravity on an object
- is measured in newtons (N).

On Earth, every kilogram of mass is pulled down with a force of 10 N.

weight (N) = mass (kg) × gravitational field strength (N/kg)

$W = m \times g$

The mass must always be in kilograms.

Terminal velocity

The force of gravity on a large mass is greater than on a small mass, but the large mass also needs a greater force to accelerate it. The two effects cancel out, and so all masses fall at the same rate in a vacuum. This acceleration due to gravity is 10 m/s^2 on the Earth.

EXAM ALERT!

More than one-fifth of students failed to get any marks at all on a recent exam question on this subject.

On Earth, air resistance affects any object moving through the air. When an object reaches terminal velocity, the air resistance is balancing the force from gravity. The air resistance does not become zero at terminal velocity!

Students have struggled with this topic in recent exams - **be prepared!**

Now try this

1. Calculate the weight of a 35 g mouse. **(3 marks)**
2. A skydiver has a weight of 700 N. She has reached terminal velocity. Explain the size of her:
   (a) acceleration **(2 marks)**
   (b) air resistance. **(2 marks)**

3. Explain how an increase in the skydiver's mass would affect her terminal velocity. **(3 marks)**

Stopping distances

It takes time for a moving car to come to a stop, and the car is still moving during this time. Understanding the factors that affect stopping distances is important for road safety.

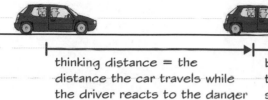

danger appears — driver brakes — car stopped

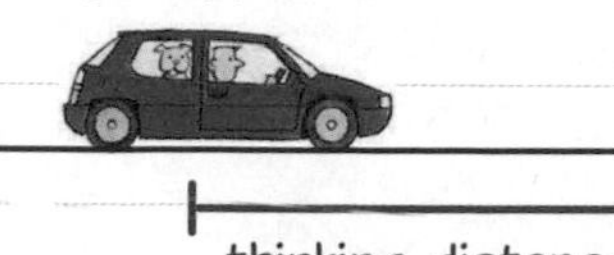

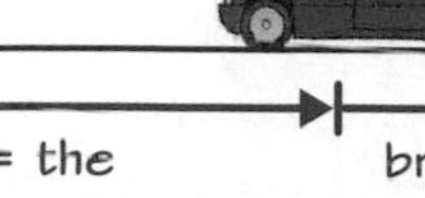

thinking distance = the distance the car travels while the driver reacts to the danger and applies the brakes

braking distance = the distance the car travels while it is slowing down

stopping distance = thinking distance + braking distance

Factors affecting stopping distance

It takes more force to stop a vehicle with more mass – so if two vehicles with the same brakes try to stop at the same time the one with the greater mass will travel further while it is braking.

Factor	Thinking distance	Braking distance
mass of vehicle	no change	*increases*
speed	*increases*	*increases*
reaction time	*increases* (if reaction time is slow)	no change
state of brakes	no change	*increases* if brakes not working properly
state of road	no change	*increases* if road surface slippery
amount of friction between tyres and road	no change	*increases* if friction is less

Reaction time is increased if the driver is tired, or has been drinking alcohol or taking drugs.

Remember that factors such as mass, brakes and road surface only affect the braking distance, not the thinking distance.

Investigating friction

You can investigate friction by using a forcemeter to pull blocks along different surfaces. You would need to keep these things the same:

- the size of the block
- the mass of the block
- the speed at which you pull it.

Now try this

1. Explain why:

(a) the driver being tired only affects the thinking distance **(3 marks)**

(b) the state of the road only affects the braking distance **(2 marks)**

(c) the speed of the vehicle affects both the thinking and the braking distances. **(4 marks)**

Momentum

The momentum of a moving object depends on its mass and its velocity. Momentum is a vector quantity.

momentum = mass × velocity
(kg m/s) (kg) (m/s)

momentum
mass × velocity

Momentum is mass multiplied by velocity, so the units for momentum are a combination of the units for mass and velocity.

Collisions

When two objects collide, the total momentum before the collision is the same as the total momentum after the collision, providing no external forces act. Momentum is conserved.

Before collision...

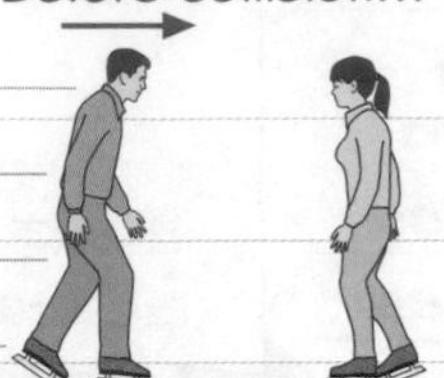

m = 50 kg
v = 5 m/s

m = 50 kg
v = 0 m/s

Total momentum
= 50 kg × 5 m/s + 50 kg × 0 m/s
= 250 kg m/s

After collision...

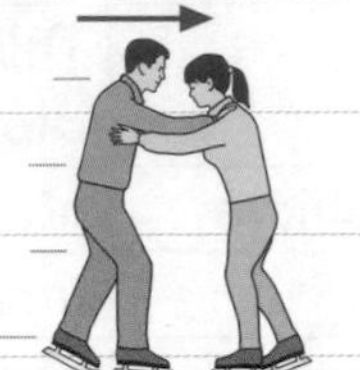

m = 100 kg
v = 2.5 m/s

Total momentum
= 100 kg × 2.5 m/s
= 250 kg m/s

The mass of the moving objects has doubled, so the velocity has halved.

Worked example

An ice skater has a mass of 65 kg. She is moving at 10 m/s when she collides with a stationary spectator standing on the ice and holds onto him. The spectator has a mass of 80 kg. Calculate the velocity of the skater and spectator together.

momentum = 65 kg × 10 m/s = 650 kg m/s

total mass = 65 kg + 80 kg
= 145 kg

Remember that the total momentum is the same before and after the collision.

$$\text{velocity} = \frac{\text{momentum}}{\text{mass}}$$

$$\text{velocity} = \frac{650\text{ kg m/s}}{145\text{ kg}}$$

= 4.48 m/s

Check your answer. The mass has gone up so the velocity must be less than the original value.

Now try this

1. A car has a mass of 1200 kg. It is moving at 15 m/s. Calculate its momentum. **(3 marks)**
2. A moving van crashes into the back of a stationary car and the two vehicles move along the road together. Explain why the velocity of the two vehicles is less than the original velocity of the van. **(3 marks)**

target C-B

3. A girl and her bicycle have a total mass of 50 kg. Her momentum is 300 kg m/s. Calculate her velocity. **(3 marks)**

Momentum and safety

Changing momentum

Momentum is conserved in collisions if there are no other forces involved. The momentum of a moving object can change if a force is applied. A larger force produces a greater rate of change of momentum (the momentum changes faster).

Car safety features

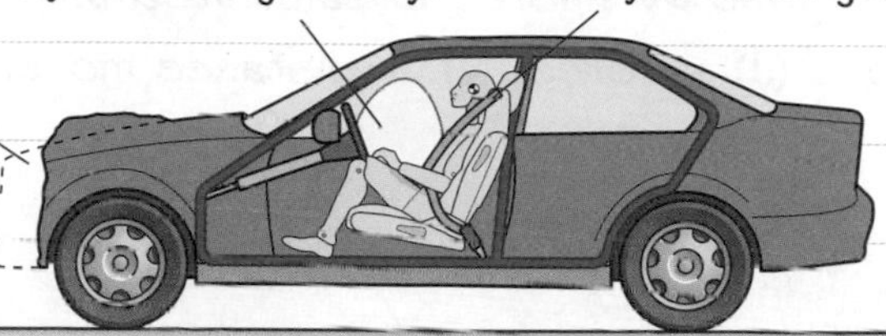
Crumple zones squash and reduce the momentum gradually.
Airbags help to slow you down gradually.
Seat belts stretch to slow you down gradually.

Bubble wrap is used to protect fragile items. If something hits the wrapped object the air in the bubbles squashes and reduces the force on the object.

Worked example

Use the idea of rate of change of momentum to explain how seat belts work.

Seat belts stretch and slow you down gradually, instead of you coming to a stop suddenly when you hit the dashboard. When you slow down more gradually the rate of change of momentum is less and so there are smaller forces on you. You will be less likely to be injured in a crash.

EXAM ALERT!

Nearly half of all students got no marks at all on a recent exam question on this subject. Remember – there are smaller forces on you if you come to a stop gradually. Smaller forces do not cause as much injury. These safety features work because they *increase the time* it takes to come to a stop.

Students have struggled with this topic in recent exams - **be prepared!** ResultsPlus

Investigating crumple zones

You can use apparatus like this to investigate the effectiveness of crumple zones. Greater impact forces will cause bigger dents in the Plasticine®.

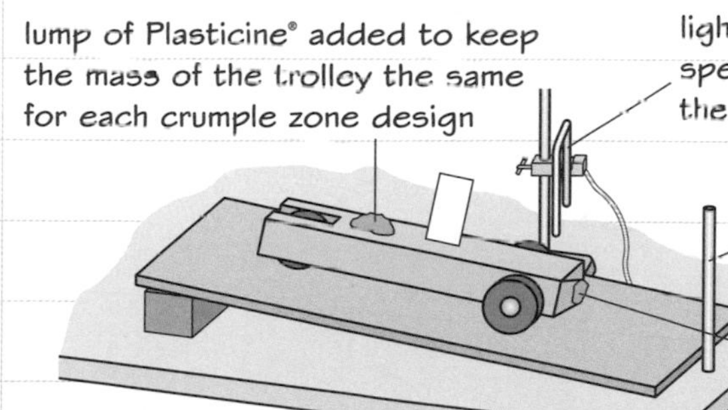
lump of Plasticine® added to keep the mass of the trolley the same for each crumple zone design
light gate measures the speed of the trolley when the card passes through it
post for the trolley to hit
model crumple zone made from card or other materials, with Plasticine® on the front

You need to keep these things the same for each crumple zone design to allow you to compare your results:

- speed of trolley
- mass of trolley
- size and shape of the object that it hits.

Now try this

target D-C

1. Explain how a seatbelt works. **(3 marks)**
2. You jump off a high wall. Explain why you should allow your knees to bend when you land. **(3 marks)**

target B-A

3. A passenger in the front seat of a car does not wish to wear the seatbelt. He says he has seen injuries caused by seatbelts. Explain why the seat belt should be worn. **(3 marks)**

Work and power

Work

Work is the amount of energy transferred, and is measured in joules (J).
The work done by a force is calculated using this formula:

work done (J) = force (N) × distance moved in the direction of the force (m)

$E = F \times d$

The E stands for energy.

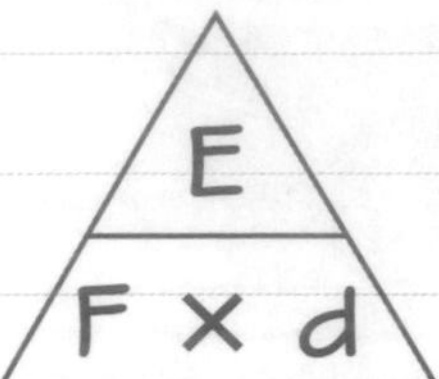

The distance moved must be in the same direction as the force.

Power

Power is the rate of doing work (how fast energy is transferred). Power is measured in watts (W). 1 watt is 1 joule of energy being transferred every second.

$$\text{power (W)} = \frac{\text{work done (J)}}{\text{time taken (s)}}$$

$$P = \frac{E}{t}$$

E
P × t

Don't get work and power mixed up. Remember:
Work = energy transferred, measured in joules.
Power = *rate* of energy transfer, measured in watts.

Worked example

Dan uses a force of 100 N to push a box across the floor. He pushes it for 3 m. Calculate the work done.

Work = 100 N × 3 m
= 300 J

Worked example

Dan takes 5 seconds to push the box across the floor. Calculate his power.

$$\text{Power} = \frac{300\text{ J}}{5\text{ s}}$$
$$= 60\text{ W}$$

Now try this

1. Joe does 60 000 J of work dragging a sledge for 10 minutes. Calculate his power.
(3 marks)

2. Jen lifts a pot of paint onto a shelf 1.5 metres above the floor. The paint weighs 10 N. It takes her 2 seconds to do this. Calculate her power.
(4 marks)

Potential and kinetic energy

Gravitational potential energy

Gravitational potential energy is the energy stored in an object because it is in a high position.

On Earth, the gravitational field strength is 10 N/kg.

gravitational potential energy (J) = mass (kg) × gravitational field strength (N/kg) × vertical height (m)

$$GPE = m \times g \times h$$

Kinetic energy

Kinetic energy is the energy stored in moving objects.

kinetic energy (J) = $\frac{1}{2}$ × mass (kg) × velocity^2 (m/s)^2

$$KE = \frac{1}{2} \times m \times v^2$$

Don't forget to square the velocity. v^2 means $v \times v$

Worked example

A 1200 kg car is travelling at 25 m/s. Calculate the kinetic energy.

$$KE = \frac{1}{2} \times 1200\,\text{kg} \times 25^2\,(\text{m/s})^2$$

$$= 375\,000\,\text{J}$$

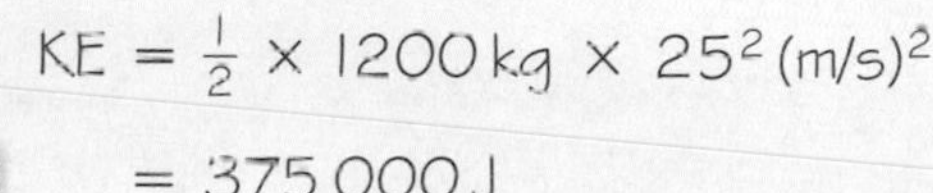

Conservation of energy

Energy cannot be created or destroyed. It can only be transferred from one form to another.

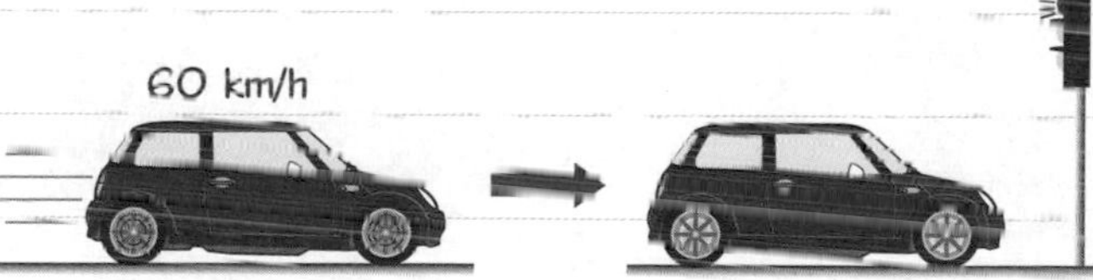

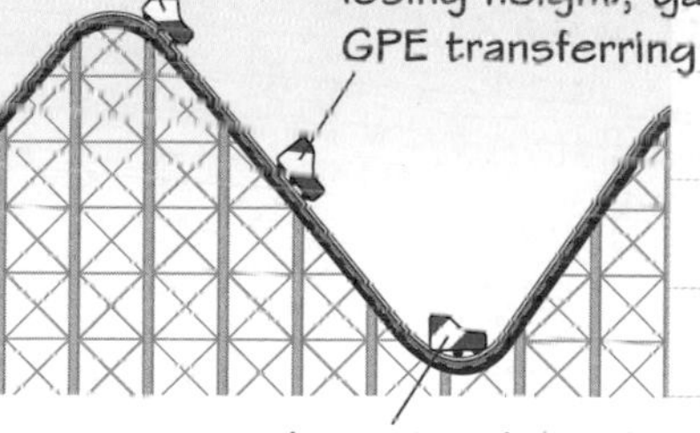

The roller coaster will eventually stop, as some of the kinetic and potential energy is transferred to thermal energy through friction and air resistance.

Now try this

target B-A

1. Anita lifts a box onto a shelf. The box has a mass of 0.5 kg, and the shelf is 2 m from the floor.
   (a) Calculate the gravitational potential energy of the box. **(3 marks)**
   (b) The box falls off the shelf. Calculate the speed of the box just before it hits the floor. **(3 marks)**

EXAM ALERT!

Over nine out of ten students did not get any marks when answering a question on this subject in a recent exam paper. Remember that the work done to move an object to a higher position is the energy transferred to it. So the work done is the same as the amount of gravitational potential energy the object has in its new position.

Students have struggled with exam questions similar to this - **be prepared!** ResultsPlus

Braking and energy calculations

Forces and momentum

A force is needed to make an object change speed. The size of the force needed depends on the change in momentum, and on how quickly this change happens.

$$\text{force (N)} = \frac{\text{change in momentum (kg m/s)}}{\text{time (s)}}$$

$$F = \frac{(mv - mu)}{t}$$

m = mass (kg)

v = final speed (m/s)

u = initial speed (m/s)

Worked example

The mass of a cyclist and his bike is 90 kg. He takes 5 seconds to slow down from 10 m/s to 2 m/s. What force are his brakes exerting?

$$F = \frac{(90\,\text{kg} \times 2\,\text{m/s} - 90\,\text{kg} \times 10\,\text{m/s})}{5\,\text{s}}$$

$$= \frac{(180 - 900)}{5}$$

$$= -144\,\text{N}$$

The minus sign shows that the force is acting in the opposite direction to the movement.

Braking distance and velocity

Worked example

A car has a mass of 1000 kg and is travelling at 10 m/s. It takes 12 metres to come to a stop.

(a) Calculate the braking force needed to do this.

$$KE = \frac{1}{2} \times 1000\,\text{kg} \times (10\,\text{m/s})^2$$

$$= 50\,000\,\text{J}$$

$$\text{force} = \frac{\text{work}}{\text{distance}}$$

$$F = \frac{50\,000\,\text{J}}{12\,\text{m}}$$

$$= 4167\,\text{N}$$

The work done to stop a car is equal to its initial kinetic energy.

(b) Calculate the braking distance if the same braking force is used from a speed of 20 m/s.

$$KE = \frac{1}{2} \times 1000\,\text{kg} \times (20\,\text{m/s})^2$$

$$= 200\,000\,\text{J}$$

$$\text{distance} = \frac{\text{work}}{\text{force}}$$

$$= \frac{200\,000\,\text{J}}{4167\,\text{N}}$$

$$= 48\,\text{m}$$

The speed has doubled, but the braking distance has gone up by a factor of 4. This shows that the braking distance depends on the velocity *squared*.

Now try this

1. A 400 g football is caught by a goalkeeper and brought to a stop in 0.5 seconds. The force on the ball is 16 N. Calculate the initial velocity of the ball. **(4 marks)**

2. A car travelling at 15 m/s comes to a stop over a distance of 25 metres. If the same braking force is used, explain what its braking distance will be from a speed of 45 m/s. **(4 marks)**

Physics extended writing 2

Worked example

A student goes for a bicycle ride. During their ride they accelerate, pedal at a steady speed, freewheel along a level road without pedalling, freewheel down a hill to gain speed, and brake hard to come to a stop quickly.

For some of these stages of their ride, it is an advantage to have a light bicycle. For others a heavy bicycle would be more useful. Explain how the mass of the bicycle affects the student at the different stages of their ride.

(6 marks)

Sample answer 1

Acceleration depends on force and mass, so if the bike is lighter the student will be able to accelerate more for the same force. Braking is a kind of acceleration, so if the bike is lighter the student will decelerate better for the same braking force.

This is a good answer. There is a good description of how the mass of the bicycle will affect acceleration and braking. However, the answer does not include anything about freewheeling along a level road or down a hill.

Sample answer 2

The student is changing their momentum when they are accelerating or braking. If their bike is lighter their momentum will be less for the same speed, and so they will accelerate faster (or come to a stop quicker) for the same force.

When they are freewheeling on a flat road, air resistance and friction are trying to slow them down, which will change their momentum. A heavy bike is better here, because it will have a bigger momentum so it will take them longer to slow down if they are not pedalling.

When they go downhill, a heavy bike is also good, because their total weight will be bigger. Their air resistance will be the same so the resultant downwards force will be a little bigger with a heavier bike. However, it is also harder to accelerate a heavier bike in the first place.

This is an excellent answer that covers all the main points. This answer discusses acceleration and braking in terms of momentum, and the first answer talked about force and mass. Either way of discussing this would earn the marks. The answer is well organised into paragraphs and uses correct scientific words.

Now try this

1. The table shows the stopping distances for a car at different speeds (measured in miles per hour).

Explain why the thinking distance is proportional to the speed, but the braking distance is proportional to speed squared. Use ideas about kinetic energy and the work done by the brakes in your answer.

(6 marks)

Speed (mph)	Thinking distance (metres)	Braking distance (metres)	Stopping distance (metres)
20	6	6	12
40	12	24	36
60	18	55	73

Physics extended writing 3

Worked example

An aeroplane with a propeller moves forwards because the spinning propeller pushes air backwards. A rocket moves forwards because fuel and oxygen carried in the rocket burn to produce hot gases that are pushed out from the rear of the rocket. Compare the way that rockets and aeroplanes move, using ideas about momentum and action and reaction forces. **(6 marks)**

Sample answer 1

The aeroplane moves because it applies an action force to the air that pushes it backwards. The reaction force is the force from the air pushing the aeroplane forwards. This force is the same size as the action force. The rocket works in a similar way, because the action force of the rocket pushing the gases is balanced by the reaction force from the gases pushing on the rocket.

This is a limited answer. It correctly describes the action and reaction forces on the aeroplane, but it states that these forces are 'balanced' on the rocket, which is not correct. The answer does not mention momentum. It is important to check the question and make sure you have included everything asked for.

Sample answer 2

Aeroplanes and rockets both work by giving momentum to gases. The propeller gives momentum to air by pushing it backwards, and the rocket engine gives momentum to the gases by pushing them backwards. Momentum is conserved, so if the gases have been given momentum there must be an equal momentum in the opposite direction on the rocket and aeroplane.

The force of the propeller or rocket on the gases is an action force. The force of the gases pushing on the propeller or rocket is the reaction force. It is this force that gives the aeroplane and rocket momentum and makes them move forwards.

They are different because the rocket pushes out gases that it has carried with it. The aeroplane uses its propeller to move the air outside it.

This is an excellent answer. It has included all the details asked for, including a comment about the differences between the two vehicles.

Now try this

1. A bouncing ball transfers gravitational potential energy to kinetic energy and back again. The efficiency of an energy transfer is the percentage of the total energy transferred that is transferred as useful energy.

   Explain why the ball does not bounce to the same height each time and how you could investigate the efficiency of a bouncing ball. **(6 marks)**

Isotopes

Atoms

Atoms are made up of protons, neutrons and electrons. The protons and neutrons are in the nucleus of the atom, and the electrons move around the outside.

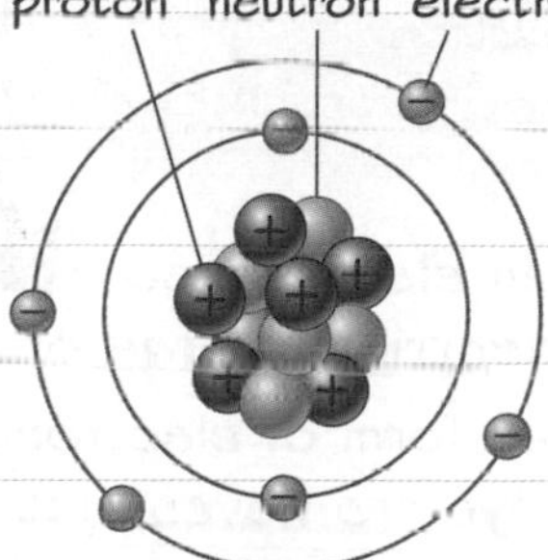

not to scale

Describing atoms

All the atoms of a particular element have the same number of protons. The number of protons in each atom of an element is called the atomic number, or proton number.

The total number of protons and neutrons in an atom is the mass number, or nucleon number.

The atomic number and mass number of an element can be shown like this:

mass number — 16; atomic number — 8: $^{16}_{8}\text{O}$

Worked example

The symbol represents an atom of sodium.

$^{23}_{11}\text{Na}$

Explain what these numbers tell you about an atom of sodium.

Remember: The mass number is always more than the atomic number (except hydrogen where both numbers are 1).

The top number in the symbol is the total number of particles in the nucleus.

The atomic number means sodium has 11 protons in the nucleus. The mass number says the number of protons plus neutrons is 23, so the number of neutrons is 23 − 11 = 12.

EXAM ALERT!

Only about half of all students answering a similar question in an exam paper recently got full marks. You need to learn the parts of an atom and how they are represented.

Students have struggled with exam questions similar to this - **be prepared!** ResultsPlus

Isotopes

Atoms of a particular element always have the same number of protons, but they can have different numbers of neutrons. Atoms with the same number of protons but different numbers of neutrons are isotopes of the same element.

Now try this

1. An atom of boron (B) has 5 protons and 6 neutrons. Show the atomic number and mass number as a symbol for the isotope. **(1 mark)**
2. These are the symbols for two isotopes of nitrogen.

   nitrogen-14 $^{14}_{7}\text{N}$ nitrogen-15 $^{15}_{7}\text{N}$

   Describe the similarities and differences between the two isotopes. **(3 marks)**

Ionising radiation

Some types of radiation can cause atoms to lose electrons and become ions. An ion is an atom that has an electrical charge because it has gained or lost electrons (see also page 41).

Alpha, beta or gamma

Some elements are radioactive. Their nuclei are unstable. This means that they may decay (change) by emitting radiation. Unstable nuclei can emit alpha, beta or gamma radiation. All these are forms of ionising radiation.

- An alpha particle is equivalent to a helium nucleus. It has two protons and two neutrons (but no electrons), so it has an electrical charge of +2.
- A beta particle is an electron, so it has a negative electrical charge of −1.
- Gamma radiation is a form of electromagnetic radiation. Gamma rays are waves, not particles. This means that they do not have a charge.

Beta particles are electrons emitted from the *nucleus* of unstable atoms. An atom that emits beta particles still has all its usual electrons moving around the nucleus.

Comparing the properties

The different types of radiation have different properties. An alpha particle is more likely to ionise an atom than a gamma ray, but a gamma ray will travel further.

paper

aluminium 3 mm thick

lead few cm thick

(α) alpha particles
- will travel a few centimetres in air
- very ionising
- can be stopped by a sheet of paper

(β) beta particles
- will travel a few metres in air
- moderately ionising
- can be stopped by aluminium 3 mm thick

(γ) gamma rays
- will travel a few kilometres in air
- weakly ionising
- need thick lead to stop them

EXAM ALERT!

You need to learn the properties of the different types of radiation. About seven out of ten students did not achieve any marks on a question on this subject in a recent exam.

Students have struggled with exam questions similar to this - **be prepared!** ResultsPlus

Worked example

Write down the three types of radiation in order of

(a) how penetrating they are, starting with the most penetrating

(a) gamma, beta, alpha

(b) their ionising power, starting with the most ionising.

(b) alpha, beta, gamma

Remember – gamma rays are the *most* penetrating and the *least* ionising.

Now try this

target D-C

1. Compare alpha and beta particles. **(4 marks)**

'Compare' questions ask you to describe similarities and differences.

2. Explain why beta particles are more penetrating than alpha particles but less ionising. **(3 marks)**

Nuclear reactions

Atoms emit alpha, beta or gamma radiation in radioactive decay. Other kinds of nuclear reaction are fission and fusion (see page 95). All nuclear reactions can be sources of energy.

Fission

In a fission reaction, a large unstable nucleus splits into two smaller ones. For example, a uranium-235 nucleus splits up when it absorbs a neutron. The fission of uranium-235 produces two daughter nuclei, two or more neutrons, and also releases energy.

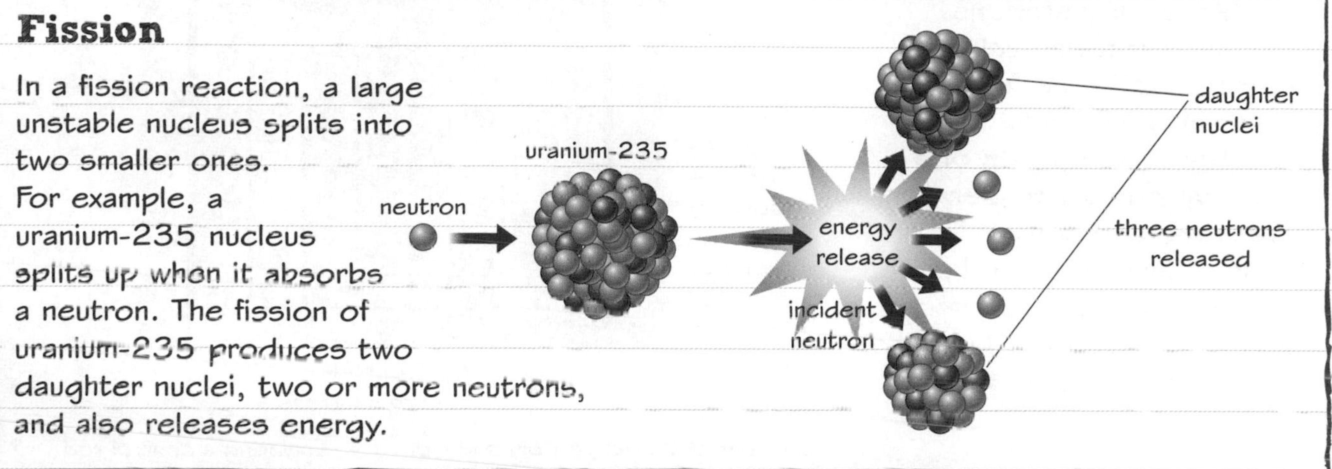

Chain reactions

The neutrons released by the fission of U-235 may be absorbed by other nuclei. Each of these nuclei may undergo fission, and produce even more neutrons. This is called a chain reaction. If a chain reaction is not controlled there will be a nuclear explosion.

Nuclear reactors make use of **controlled** chain reactions (see next page).

Worked example

Describe how a chain reaction can be controlled.

A chain reaction can be controlled by using a different material to absorb some of the neutrons. This slows the reaction down because there are fewer neutrons to cause more nuclei to fission.

Controlled chain reactions

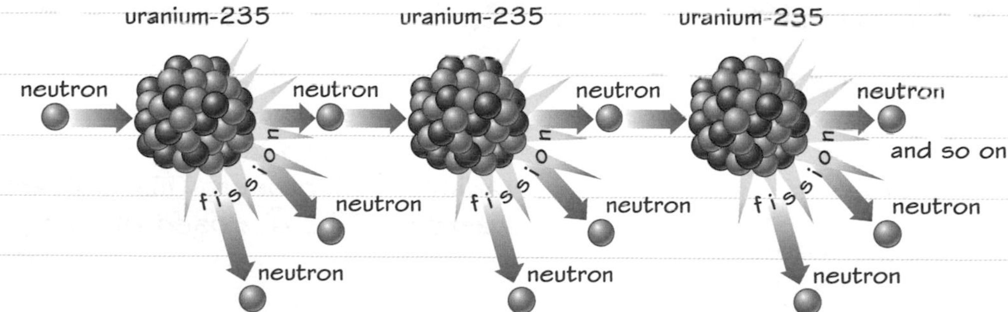

Two of the neutrons are absorbed by other materials. Only one neutron from each fission can cause other fission. This is a **controlled chain reaction**.

Now try this

1. Describe what happens in a chain reaction. **(3 marks)**
2. Explain the difference between a chain reaction and a controlled chain reaction. **(4 marks)**

Nuclear power

Nuclear power stations

Nuclear power stations use nuclear fuels such as uranium-235. The fuel is made into fuel rods. A reactor core is made of a material called a moderator. Fuel rods and control rods fit into holes in the moderator. The moderator and the control rods help to control the chain reaction.

fuel rods
control rods
moderator
reactor core
Concrete shielding prevents radiation and stray neutrons escaping from the core.
Control rods absorb neutrons.
Neutrons that escape from one fuel rod can be absorbed by another.
Lowering a control rod reduces fission reactions.

Controlling the reaction

The neutrons produced by fission reactions are moving very fast. The moderator slows them down so that they are more likely to be absorbed by another U-235 nucleus and cause another fission reaction.

The moderator does not slow down or speed up the rate of the chain reaction. The control rods are moved in or out to do that.

Worked example

Explain how the control rods work in a nuclear power station.

The control rods absorb neutrons. If the control rods are pushed down into the core, more neutrons are absorbed and the chain reaction slows down. If they are pulled out, fewer neutrons are absorbed and the chain reaction speeds up.

Generating electricity

Thermal (heat) energy released by the chain reaction is used to turn water into steam. The steam makes a turbine spin, and the turbine drives a generator.

Nuclear power stations produce steam and use it to turn turbines in the same way as fossil-fuelled power stations do. The difference is the way the heat is produced.

Radioactive waste

The daughter nuclei produced in the fission reaction are radioactive. The neutrons passing through the core can form other radioactive isotopes. This radioactive waste must be disposed of safely.

Now try this

1. Explain the purpose of the moderator in a nuclear reactor. **(3 marks)**

2. Explain how pulling control rods out of the core of a nuclear reactor can increase the amount of heat energy released by the reactor. **(4 marks)**

Fusion – our future?

Fusion

Nuclear fusion happens when small nuclei join to form larger ones. Like all nuclear reactions, fusion reactions release energy.

Isotopes of hydrogen combine in the Sun to form helium. The energy released by these reactions is what makes the Sun shine.

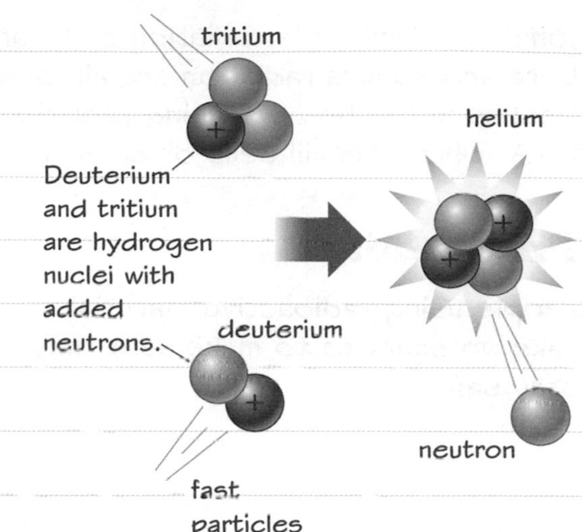

EXAM ALERT!

Remember that it is nuclei that combine in fusion reactions, not atoms. In a recent exam question on this subject, nearly half the students made this mistake.

Students have struggled with exam questions similar to this - **be prepared!** 

Difficulties of fusion

Nuclei need to get very close to each other before fusion can happen. Under normal conditions the positive charges on nuclei repel each other. Only at very high temperatures and pressures are the nuclei moving fast enough for them to overcome this repulsion.

The very high temperatures and pressures needed are very difficult to produce in a fusion power station. All the experimental fusion reactors built so far have used more energy than they have produced!

'Cold fusion'

In 1989 two scientists announced that they had made nuclear fusion happen at 50°C. This became known as 'cold fusion'. Their results have not been accepted because they could not be validated.

Scientists make a new discovery. → They describe their method and results in a paper, which they send to an academic journal. → The editor sends the paper to other scientists for checking. This is called peer review. → The journal publishes the paper, and other scientists try to reproduce the results. → The discovery is validated if the results can be reproduced.

The scientific process.

Now try this

1. Describe the difference between fission and fusion. **(2 marks)**
2. State two things that need to happen before a new scientific theory is accepted by the scientific community. **(2 marks)**
3. Explain why there are no commercial power stations using fusion reactions. **(2 marks)**

Changing ideas

Ionising radiation is radiation that can remove electrons from atoms and form ions. Alpha, beta and gamma radiation are all forms of ionising radiation. Ionising radiation can damage tissues in the body, causing radiation burns (reddened skin), or may cause mutations in DNA, which can kill cells or cause cancer.

Precautions

People using radioactive material take precautions to make sure they stay safe.

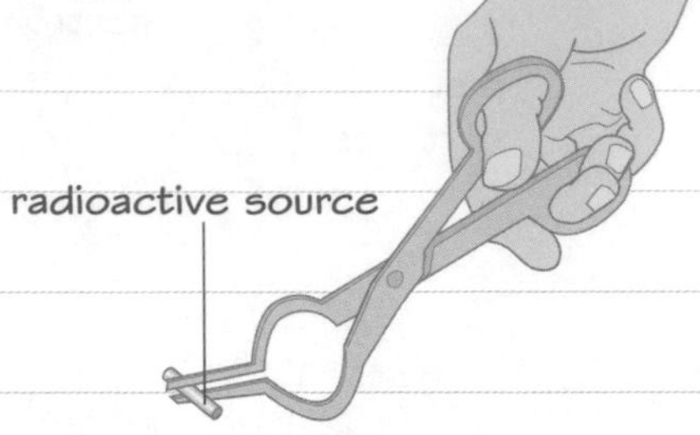

The radioactive source is being moved using tongs to keep it as far away from the person's hand as possible. The source is always kept pointing away from people.

Worked example

Explain two precautions that a radiation worker could take.

They could wear overalls to make sure radioactive particles do not get caught in their clothing.

They could wear breathing apparatus to make sure radioactive material does not get into their lungs.

Remember that if a question asks you to 'explain' something, you need to give a reason why something happens or something is done.

Changing ideas

When radioactivity was first discovered, scientists did not realise that it was harmful. Scientists using radioactivity suffered burns to their skin, but they did not realise that ionising radiation could cause cancer.	In the 1920s scientists began to see a link between working with radioactivity and the chances of getting cancer. Scientists also now know about DNA and that some changes to DNA can be caused by ionising radiation.	Today scientists have discovered how to work with radioactivity safely, including the amount of radioactivity we can be exposed to without danger.

If an exam question asks you to 'suggest' something, it means you may not have learned the answer directly. You need to use your scientific knowledge to help you to work out the answer.

Worked example

Suggest two reasons why ideas about the hazards of radioactivity have changed.

Scientists have gathered evidence about exposure to radioactivity and illness, and scientists have carried out experiments to find out *how* radioactivity is harmful.

Now try this

1. Explain two precautions that a teacher using a radioactive source in class should take. **(4 marks)**

2. A source of alpha radiation and a source of beta radiation are pointed towards a person from 1 metre away. Explain which source is the most dangerous. **(3 marks)**

Nuclear waste

Waste from nuclear power stations is radioactive and must be disposed of safely. Hospitals and research laboratories can also produce some radioactive waste. There are different ways of disposing of waste, depending on how radioactive it is.

Types of waste

- High level waste (HLW) is very radioactive. It is sealed into glass and stored until it has become less radioactive (after about 50 years, when it becomes ILW).
- Intermediate level waste (ILW) remains radioactive for tens of thousands of years. It is stored in steel and concrete containers.
- Low level waste (LLW) includes waste from hospitals, and clothing and cleaning materials from nuclear power stations. It is buried in special landfill sites.

Worked example

Suggest two ways of disposing of high and intermediate level waste. State one disadvantage of each method.

It could be fired into space, but radioactive material would be spread over a wide area if there was an accident during launch.

It could be dumped at sea, but radioactive materials could get into the oceans if the storage containers leaked.

Nuclear waste could also be stored underground. This would have to be done in an area that is not likely to have earthquakes.

Nuclear power

There are advantages and disadvantages to using nuclear energy to generate electricity.

- ✓ Nuclear power stations do not produce carbon dioxide, so they do not contribute to climate change.
- ✓ Supplies of nuclear fuel will last longer than supplies of fossil fuels.
- ✗ It is difficult and expensive to store nuclear waste safely.
- ✗ An accident in a nuclear power station can spread radioactive material over a large area.
- ✗ Many people think that nuclear power is dangerous, and do not want new nuclear power stations to be built.

However, construction processes produce carbon dioxide, so carbon dioxide will be added to the atmosphere when the power station is built and when fuel rods are made.

Nuclear power stations do not make the local area more radioactive when they are working properly. This only happens if there is an accident.

Now try this

1. 'Nuclear power does **not** produce carbon dioxide emissions.' Explain why this statement is not correct. **(3 marks)**
2. Suggest why low level nuclear waste is easier to dispose of than intermediate or high level waste. **(2 marks)**

Half-life

The activity of a radioactive source is the number of atoms that decay every second. The unit for activity is the becquerel (Bq). When an atom decays it emits radiation but changes into a more stable isotope.

Unstable atoms

The activity of a source depends on how many unstable atoms there are in a sample, and on the particular isotope. As more and more atoms in a sample decay, there are fewer unstable ones left, so the activity decreases. The half-life of a radioactive isotope is the time it takes for half of the unstable atoms to decay. This is also the time for the activity to go down by half.

Be careful when you are writing about half-life. It is not the time for an atom to decay – it is the time for half of the atoms *in a sample* to decay.

Worked example

The half-life of caesium-137 is 30 years. How long does it take for the activity of a sample to change from 100 Bq to 25 Bq?

After 30 years (1 half-life) the activity will be 50 Bq.

After 60 years (2 half-lives) the activity will be 25 Bq.

It will take 60 years.

Worked example

The graph shows how the activity of a sample changes over 24 hours. What is the half-life of the sample?

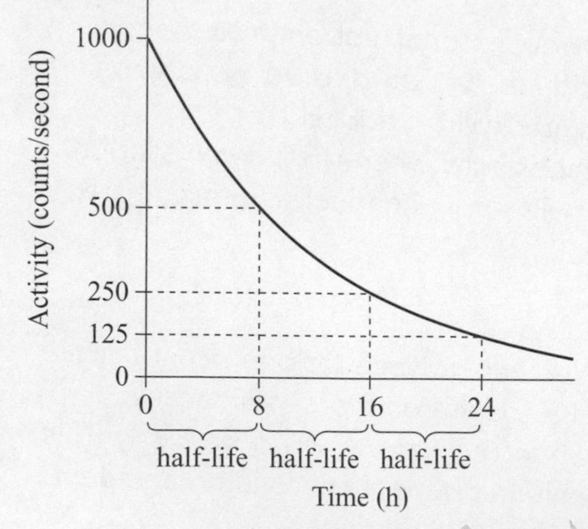

Activity at time 0 = 1000 Bq

Half of this is 500 Bq. The activity is 500 Bq at 8 hours.

The half-life is 8 hours.

EXAM ALERT!

About two-fifths of all students did not achieve full marks on a similar question in a recent exam. Remember that the half-life is the amount of time it takes for the activity to halve.

Students have struggled with exam questions similar to this - **be prepared!** ResultsPlus

Models for decay

Radioactive decay is a random process. You can model how the atoms in a sample decay using other random processes, such as flipping a coin or throwing a die.

Now try this

1. The activity of a source is 60 Bq. The activity is 15 Bq four hours later. What is the half-life of the source? **(3 marks)**

2. A source has an activity of 160 Bq and a half-life of 3 minutes. What is the activity after 15 minutes? **(3 marks)**

Uses of radiation

Background radiation

We are always exposed to ionising radiation. This is called background radiation. This radiation comes from different sources, as shown in the pie chart.

Radon is a radioactive gas that is produced when uranium in rocks decays. The radon can build up in houses and other buildings. The amount of radon gas varies from place to place, because it depends on the type of rock in the area.

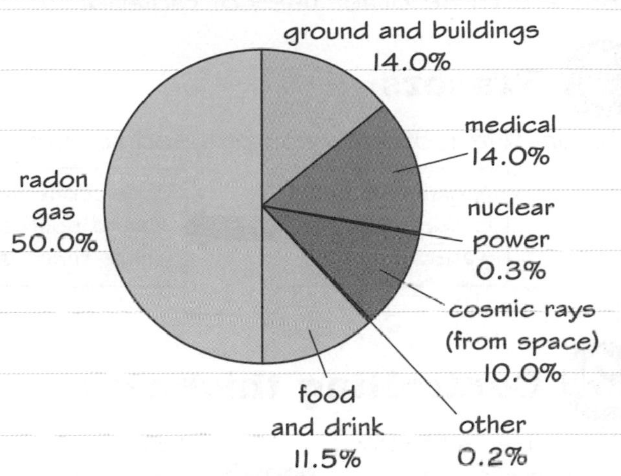

Radiation in hospitals

Radiation is used in hospitals to:

- Kill cancer cells – beams of gamma rays can be directed at cancer cells to kill them.
- Sterilize surgical instruments – gamma rays can be used to sterilize plastic instruments which cannot be sterilized by heating.
- Diagnose cancer – a tracer solution containing gamma radiation is injected into the body and taken up by cells which are growing abnormally. The places in the body where the tracer collects are detected with a 'gamma camera'.

Irradiating food

Worked example

Explain why gamma rays are used to irradiate food.

It makes the food safer to eat, and makes it last longer.

Don't get 'irradiated' and 'radioactive' mixed up. Irradiated food is not radioactive.

Bacteria on food eventually causes it to **decompose**, and some bacteria may cause food poisoning. **Irradiating** the food with gamma rays kills bacteria and any other organisms in it. This makes the food safer to eat. The food can also be stored for longer before it goes off.

Now try this

1. Explain how gamma rays can be used to:
   (a) detect cancer **(3 marks)**
   (b) treat cancer. **(2 marks)**

2. The pie-chart shows the average background radiation in the UK. Explain how radon contributes to the variation in background radiation between different parts of the UK. **(3 marks)**

More uses of radiation

Here are three other uses of radiation.

1 Tracers

Radioactive isotopes can be used to find leaks in pipes.

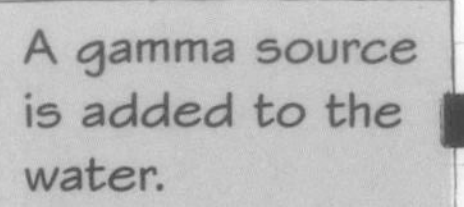

A gamma source is added to the water. → Water containing the gamma source escapes where there is a leak. → A detector above the ground detects higher levels of gamma radiation.

2 Controlling thickness

If the paper is too thick, not as many beta particles get through. → The rollers press together harder to make the paper thinner.

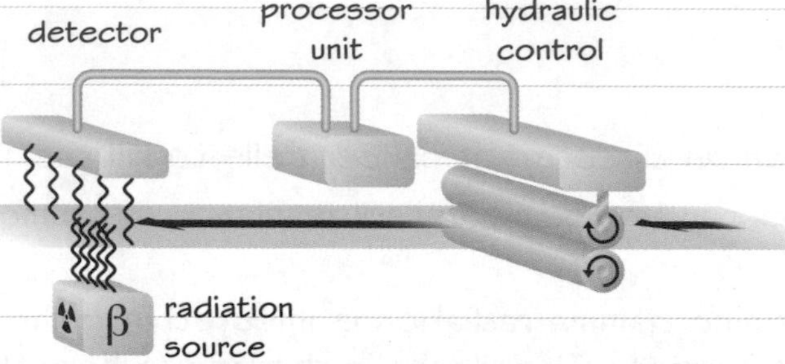
detector
processor unit
hydraulic control
β
radiation source
Beta particles being used to control the thickness of paper.

Worked example

Explain why beta particles are used to control the thickness of paper.

Alpha particles would not go through the paper at all. Gamma rays would pass through the paper too easily, and the amount getting through would hardly change with small changes in the thickness of the paper.

3 Smoke alarms

Smoke alarms contain a source of alpha radiation.

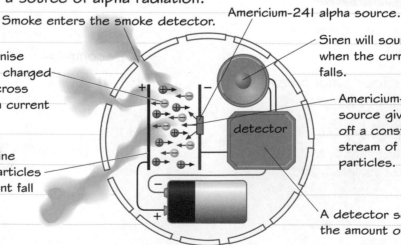

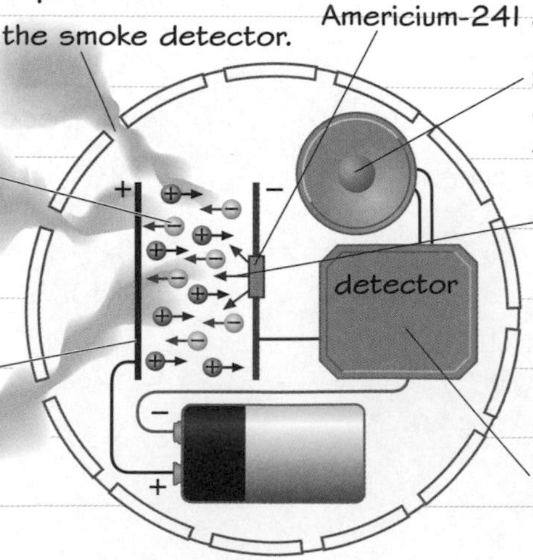
Smoke enters the smoke detector.
Americium-241 alpha source.
Siren will sound when the current falls.
Alpha particles ionise the air and these charged particles move across the gap forming a current
Americium-241 source gives off a constant stream of alpha particles.
detector
Smoke in the machine will absorb alpha particles and make the current fall
A detector senses the amount of current.

Now try this

target C-B

1. Look at the diagram showing paper being made. Describe what will happen if the paper is too thin. **(2 marks)**
2. Explain how a smoke detector works. **(4 marks)**

Physics extended writing 4

Worked example

We can be exposed to radiation from radioactive isotopes inside the body if they have been breathed in or taken in with food. We can also be exposed to radiation from outside the body.

Explain why alpha radiation is more dangerous if the source is inside the body, and beta and gamma radiation are more dangerous if the source is outside the body.

(6 marks)

Sample answer 1

Alpha radiation is more ionising so it will damage cells by knocking off electrons to form ions, which is bad because it can cause cancer so this is why alpha is more dangerous.

This is a basic answer. It contains some correct details about ionisation, but it does not mention penetrating power and it does not mention beta and gamma radiation. The answer would also be better if split up into more sentences.

Sample answer 2

Radiation can cause cancer if it ionises the DNA inside cells and causes mutations. Alpha radiation is the most ionising, so if alpha radiation hits a cell it is more likely to damage the DNA than beta or gamma radiation. However because it is the most ionising, alpha radiation is also the least penetrating. If a source of alpha radiation is outside the body, the alpha particles will be stopped by a few centimetres of air between the source and the body, or will be stopped by the skin. Alpha radiation only causes serious damage if the source is inside the body.

Beta radiation is less ionising than alpha radiation, so it can penetrate more materials. Gamma radiation is even less ionising and more penetrating. If a beta or gamma source is inside the body, the radiation may pass out of the body without damaging any cells. But this also means that beta and gamma radiation from sources outside the body can easily get into it. This is why they are more dangerous outside the body.

This is an excellent answer that covers all the main points and uses correct scientific language. The answer is also divided up sensibly into two paragraphs and uses correct spelling and grammar.

Now try this

1. Radioactive decay can be modelled by tossing a coin or throwing dice. Compare how these two methods model radioactive decay. **(6 marks)**

Physics extended writing 5

Worked example

'Nuclear power stations increase the amount of radiation we are exposed to.'

Discuss this statement. **(6 marks)**

'Discuss' means that your answer needs to reach a conclusion about whether or not you agree with the statement.

Sample answer 1

Nuclear power is very dangerous because it produces radioactive waste that is radioactive for thousands of years, and sometimes nuclear power stations can explode and the radiation can get everywhere. The statement is correct.

This is a basic answer. The information in the answer is more or less correct, although it does not take into account the fact that nuclear accidents are rare, and nuclear waste is stored so that radioactive materials do not enter the environment. The answer would be improved by explaining other sources of radiation.

Sample answer 2

We are exposed to background radiation all the time. In some parts of the country most of this comes from radioactive radon gas released by rocks. Background radiation also comes from space, from our food and drink, and also from radioactive isotopes used in hospitals.

Nuclear power stations are responsible for a very small proportion of background radiation when they are working normally. They only release a lot of radiation if there is an accident, and this does not happen very often.

So I think the statement is correct, but the amount of the increase is very small.

This is an excellent answer. It discusses the amount of radiation from nuclear power stations in comparison to normal background radiation, and draws a conclusion about the statement in the question.

Now try this

1. A hospital uses radioactive isotopes:
   - to inject into cancer tumours to kill the cancer cells
   - to inject into the body to use as a tracer in diagnosis
   - to focus gamma rays at a tumour from outside a patient.

Isotope	Half-life	Main type of radiation emitted
actinium-225	10 days	alpha
cobalt-60	1925 days	gamma
poloniuim-216	0.15 seconds	alpha
technetium-99m	6 hours	gamma

The table shows some characteristics of some different radioactive isotopes. Explain which of these isotopes would be best for each use.

(6 marks)

Balancing equations

You can represent what happens in a chemical reaction using equations. Equations can be in words or they can use symbols for the elements and compounds.

If you are asked for a word equation, don't be tempted to use the formulae as a short-cut.

hydrogen + oxygen → water

$2H_2 + O_2 \rightarrow 2H_2O$

There are 2 oxygen atoms here as well, because there are 2 molecules of water, each with 1 oxygen atom.

There are 2 oxygen atoms here, joined into 1 molecule.

This is a **balanced equation**: it has the same number of hydrogen and oxygen atoms on each side of the → sign.

Remember:

- Count the number of different atoms on each side of the → sign to check if your equation is balanced.
- The big number in front of a formula tells you how many molecules of a covalent substance, units of an ionic substance or atoms of a metal there are.
- You must not try to balance an equation by changing the formulae of the compounds! You can only add more molecules.
- You may be asked to include state symbols in your equation – (s) for solid, (l) for liquid, (g) for gas and (aq) for a substance in solution.

Worked example

Lithium (Li) reacts with water to form lithium hydroxide (LiOH) and hydrogen gas. Write a balanced equation to represent this reaction. Include state symbols.

You are expected to know the symbols and formulae for some of the common substances, such as water and hydrogen gas.

$2Li(s) + 2H_2O(l) \rightarrow 2LiOH(aq) + H_2(g)$

Start by writing down all the formulae and state symbols:

$Li(s) + H_2O(l) \rightarrow LiOH(aq) + H_2(g)$

Count the atoms on each side. You can easily see there are more hydrogen atoms on the right, so you need to add more on the left. You can only do this by putting a 2 in front of the water.

$Li(s) + 2H_2O(l) \rightarrow LiOH(aq) + H_2(g)$

Now there is more oxygen on the left than on the right, so increase the number on the right by putting a 2 in front of LiOH.

$Li(s) + 2H_2O(l) \rightarrow 2LiOH(aq) + H_2(g)$

Now you need to increase the number of lithium atoms on the left, so put a 2 in front of Li.

$2Li(s) + 2H_2O(l) \rightarrow 2LiOH(aq) + H_2(g)$

Count the atoms on each side again to check – the equation is now balanced.

Practical work

The Edexcel Additional Science course includes suggestions for many different investigations. By the time you sit your exams you will have completed a Controlled Assessment, based on one or more of these investigations. But you could also be asked questions about any of these practicals in the exam.

This revision guide includes a brief summary of a method that could be used for each of the suggested investigations. The worked examples below should help with other kinds of questions.

Questions based on practical work could include:

- providing a hypothesis, and justifying it
- writing a method for an investigation
- explaining how to control variables
- drawing a graph to show some results
- writing a conclusion based on results given in the exam paper
- evaluating a method or a conclusion.

Worked example

A student is investigating the relationship between potential difference, current and resistance. Write a hypothesis and method for this investigation.

Hypothesis: if the resistance in a circuit is increased, the current will be smaller for a particular potential difference. This is because a higher resistance means it is harder for a current to flow through the component.

This is a good hypothesis, because it shows that the student has recalled what they have learned about current, potential difference and resistance.

Method: Set up a circuit with an ammeter in series with a resistor, and with a power pack or battery to provide a constant potential difference. Measure the current. Add another resistor to the circuit and measure the current again. Keep doing this until the current has been measured for 5 different resistances of the circuit.

You should usually have at least 5 data points if you want to draw a graph of your results.

Plot a graph of resistance against current to see the relationship between the two variables.

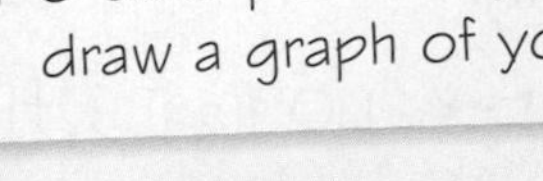

You must describe a method in the correct order. It may help to jot down some ideas in a blank space on your exam paper to help you to get your ideas in order. If the question asks you to explain the method, remember to say *why* each step is needed.

Other questions on planning

Other questions on the planning part of a practical could include asking you to:

- explain the apparatus needed
- identify the variables to be controlled and explain how they can be controlled
- identify risks and describe how to manage them.

Dealing with evidence

Worked example

The graph shows the results of an investigation to find out how the rate of anaerobic respiration in yeast depends on the concentration of glucose. The rate of respiration was measured by finding the volume of carbon dioxide produced in 10 minutes. Complete the graph by drawing a line of best fit.

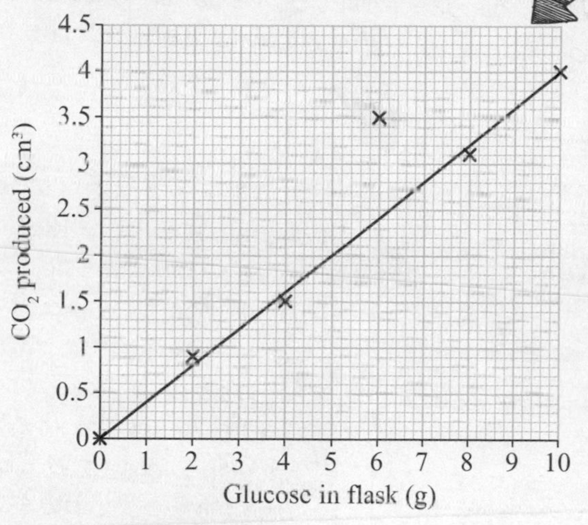

A 'line of best fit' can be a straight line or it can be a smooth curve. In this question, the points are along a straight line (apart from the one that obviously does not fit the pattern), so draw a straight line using a ruler.

This is a good answer because the student has drawn a single straight line through most of the points. You will not get any marks for a graph question if you join each of the points using straight lines. Try to include as many points as possible on your line or curve, but ignore any that are obviously not following the pattern of most of your results.

A result that does not fit the pattern is an **anomalous result**. The student has probably made a mistake when measuring or recording the result for 6 g of glucose.

Do not include anomalous results when you are working out means, or when you are drawing lines or curves of best fit.

Conclusions and evaluations

Worked example

A student investigating respiration in yeast had the following hypothesis: 'The rate of respiration will increase when the concentration of glucose increases.'

Look at the graph of their results (above).

(a) Write a conclusion for this investigation.

(a) The graph shows that the rate of respiration of the yeast increases when the concentration of glucose increases. The hypothesis was correct.

This would be a better answer if the conclusion described the shape of the graph in more detail. The graph is a straight line through the origin (0, 0), so the volume of CO_2 produced is proportional to the mass of glucose used. Proportional means that the CO_2 doubles if the glucose doubles.

(b) Evaluate the quality of the conclusion.

(b) All the points except one are close to the line of best fit. The result for 6 g of glucose was probably a mistake. The quality of the data could be improved by using several flasks with each mass of glucose and finding means (averages) of the results.

The conclusion may not apply if more glucose is used. Too much glucose might be bad for the yeast.

Many relationships are only true up to certain values of one of the variables. So you can only really say that your conclusion is valid for the range of the variables you have tested.

Final comments

Here are some other things to remember in your exam.

Read the question carefully. Underline important words in the question to help you to understand what you need to do. Using correct science is great, but no use if it does not actually answer the question!

Use the correct scientific words for things and don't be vague in your answers. For example, saying 'fossil fuels cause pollution' isn't specific enough. Saying *how* they cause pollution is better ('burning fossil fuels causes pollution because carbon dioxide and sulfur dioxide get into the air').

Know what the command words mean at the start of a question. If you are asked to 'explain' then you need to say what happens and how or why it happens. If a question asks you to 'compare' then you need to write down something about all the things you are comparing and how they are similar and different.

Write your name here
Surname | Other names
Centre Number | Candidate Number
Edexcel GCSE
Chemistry/Additional Science
Unit C2: Discovering Chemistry
Higher Tier
Time: 1 hour
Paper Reference **XXX**
You must have: Calculator, ruler
Total Marks

Instructions

- Use **black** ink or ball-point pen.
- **Fill in the boxes** at the top of this page with your name, centre number and candidate number.
- Answer **all** questions.
- Answer the questions in the spaces provided – *there may be more space than you need.*

Information

- The total mark for this paper is 60.
- The marks for **each** question are shown in brackets – *use this as a guide as to how much time to spend on each question.*
- Questions labelled with an **asterisk** (*) are ones where the quality of your written communication will be assessed – *you should take particular care with your spelling, punctuation and grammar, as well as the clarity of expression, on these questions.*

Advice

- Read each question carefully before you start to answer it.
- Keep an eye on the time.
- Try to answer every question.
- Check your answers if you have time at the end.

Turn over

P40176A

edexcel
advancing learning, changing lives

Don't use a formula for chemical substances unless the question asks for it. For example, if your answer is carbon dioxide and you write C_2O by mistake (because carbon dioxide is CO_2), it won't be possible to tell that you knew the right answer.

Learn how to balance chemical equations. And remember that all the gases that take part in the reactions you need to know about for this course are diatomic (their formulae are: O_2, Cl_2, N_2, and so on).

Revise the investigations you have carried out during the course. You may be asked questions on practical work in the exam.

Show your working for calculation questions, even if you use a calculator. And don't forget to include the units with your final answer.

Using formula triangles

There will be a formula sheet in the exam, so you do not need to memorise equations, but you do need to be able to rearrange them.

If you cannot remember how to do this, you need to memorise the formula triangles given with formulae in this book. For example, $P = I \times V$ will be given in the exam paper. If you need to work out the voltage (V), cover up the V on the formula triangle. This will tell you that you need to divide P by I to get your answer.

$P = I \times V$ (given in exam)

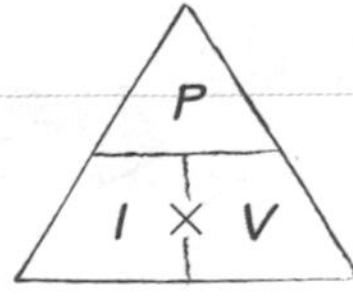

This can be rearranged as:

$V = \frac{P}{I}$ or $I = \frac{P}{V}$

Periodic table

You will be given a copy of the periodic table in your exam. It will look something like this.

Key

relative atomic mass **atomic symbol** name atomic (proton) number

1 **H** hydrogen 1

1	2											3	4	5	6	7	0
																	4 **He** helium 2
7 **Li** lithium 3	9 **Be** beryllium 4											11 **B** boron 5	12 **C** carbon 6	14 **N** nitrogen 7	16 **O** oxygen 8	19 **F** fluorine 9	20 **Ne** neon 10
23 **Na** sodium 11	24 **Mg** magnesium 12											27 **Al** aluminium 13	28 **Si** silicon 14	29 **P** phosphorus 15	31 **S** sulfur 16	35.5 **Cl** chlorine 17	40 **Ar** argon 18
39 **K** potassium 19	40 **Ca** calcium 20	45 **Sc** scandium 21	48 **Ti** titanium 22	51 **V** vanadium 23	52 **Cr** chromium 24	55 **Mn** manganese 25	56 **Fe** iron 26	59 **Co** cobalt 27	59 **Ni** nickel 28	63.5 **Cu** copper 29	65 **Zn** zinc 30	70 **Ga** gallium 31	73 **Ge** germanium 32	75 **As** arsenic 33	79 **Se** selenium 34	80 **Br** bromine 35	84 **Kr** krypton 36
85 **Rb** rubidium 37	88 **Sr** strontium 38	89 **Y** yttrium 39	91 **Zr** zirconium 40	93 **Nb** niobium 41	96 **Mo** molybdenum 42	[98] **Tc** technetium 43	101 **Ru** ruthenium 44	103 **Rh** rhodium 45	106 **Pd** palladium 46	108 **Ag** silver 47	112 **Cd** cadmium 48	115 **In** indium 49	119 **Sn** tin 50	122 **Sb** antimony 51	128 **Te** tellurium 52	127 **I** iodine 53	131 **Xe** xenon 54
133 **Cs** caesium 55	137 **Ba** barium 56	139 **La*** lanthanum 57	178 **Hf** hafnium 72	181 **Ta** tantalum 73	184 **W** tungsten 74	186 **Re** rhenium 75	190 **Os** osmium 76	192 **Ir** iridium 77	195 **Pt** platinum 78	197 **Au** gold 79	201 **Hg** mercury 80	204 **Tl** thallium 81	207 **Pb** lead 82	209 **Bi** bismuth 83	[209] **Po** polonium 84	[210] **At** astatine 85	[222] **Rn** radon 86
[223] **Mn** francium 87	[226] **Ra** radium 88	[227] **Ac*** actinium 89	[261] **Rf** rutherfordium 104	[262] **Db** dubnium 105	[266] **Sg** seaborgium 106	[264] **Bh** bohrium 107	[277] **Hs** hassium 108	[268] **Mt** meitnerium 109	[271] **Ds** darmstadtium 110	[272] **Rg** roentgenium 111	Elements with atomic numbers 112-116 have been reported but not fully authenicated						

** The lanthanoids (atomic numbers 58-71) and the actinoids (atomic numbers 90-103) have been omitted.*

The relevant atomic masses of copper and chlorine have not been rounded to the nearest whole number.

Answers

You will find some advice next to some of the answers. This is written in italics. It is not part of the mark scheme but just gives you a little more information.

3. Plant and animal cells

1.

Component	Found in plant cells?	Found in animal cells?	Function
cell wall	yes	no	gives cell a rigid shape
cell membrane	yes	yes	controls what enters and leaves cell
chloroplasts	yes	no	where photosynthesis takes place and makes food
cytoplasm	yes	yes	to support other structures and where some reactions happen
mitochondria	yes	yes	where respiration releases energy
nucleus	yes	yes	contains DNA, which controls what is made and how the cell works
vacuole	yes	no	contains cell sap and helps to support plant when full

(1 mark for each row)

2. The plant is supported by the cell wall around each cell **(1)** and the vacuole in each cell when it is full. **(1)**
3. Not all plant cells photosynthesise, e.g. root hair cells. **(1)** Cells that don't photosynthesise don't need chloroplasts. **(1)**

4. Inside bacteria

1. 1.2/(40 × 10) **(1)** = 0.003 mm / 3×10^{-6} m/ 3 μm **(1)** *You can give any of these units for your answer – but unless you are very sure that you can convert the units properly, it is best to stick to the answer you get from the units used in the question (mm, in this case). Unless, of course, the question asks for the answers in a different unit.*
2. Chromosomal DNA forms a long loop that lies free in the cytoplasm and contains most of the genes; **(1)** plasmid DNA forms small circles of DNA in the cytoplasm, which carry additional genes. **(1)**

5. DNA

1. Any suitable sentences that give appropriate definitions, such as: A gene is a short section of DNA that codes for a specific protein; **(1)** there are four bases in DNA: A, T, C and G. **(1)**
2. GCTA **(1)**; G always pairs with C and T always pairs with A. **(1)**
3. DNA forms a double helix structure **(1)**, which consists of two strands coiled together **(1)** and held together by the weak hydrogen bonds **(1)** that link the complementary base pairs of A and T or C and G. **(1)**

6. DNA discovery

1. They each contributed some of the work that led to the final working out of the structure. **(1)**
2. When many scientists collaborate they can either complete the work faster (as in the decoding of the human genome) **(1)** or contribute different bits of information from different techniques (as in the discovery of the DNA structure). **(1)**

7. Genetic engineering

1. An organism that has a gene inserted into its DNA **(1)** that comes from another organism. **(1)**
2. Genes from other plants that code for beta-carotene are cut out of/ isolated from the DNA of those plants; **(1)** the genes are inserted into the DNA of rice plants **(1)** so that the rice plants produce beta-carotene. **(1)**
3. Any two from: could reduce biodiversity; **(1)** could harm human health when eaten; **(1)** characteristic could be transferred to other plants. **(1)** You will get no marks for saying that some people think it is wrong to change the genes of organisms, because the question asks for **scientific** arguments.

8. Mitosis

1. **(a)** A cell that has two sets of chromosomes. **(1)**
   (b) A cell produced by division of a parent cell. **(1)**
2. When the cell divides each daughter cell gets one of the copies of each chromosome; **(1)** each chromosome is copied exactly **(1)** so both daughter cells have identical chromosomes. **(1)**
3. They will be the same colour **(1)** because the cells in the plants are genetically identical. **(1)**

9. Fertilisation and meiosis

1. Any correct use of each word, such as: During fertilisation, a male haploid gamete and a female haploid gamete combine to form a diploid zygote. (**max. 4**)
2. Any two comparisons from: 2 daughter cells produced in mitosis, **(1)** 4 daughter cells produced in meiosis; **(1)** daughter cells get 2 sets of chromosomes/are diploid in mitosis, **(1)** daughter cells get 1 set of chromosomes/are haploid in meiosis; **(1)** daughter cells are genetically identical in mitosis, **(1)** daughter cells are genetically different in meiosis; **(1)** mitosis used for growth, repair and sexual reproduction, **(1)** meiosis is used to produce gametes. **(1)**
3. Meiosis reduces the chromosome number to half the diploid number; **(1)** during fertilisation two gametes fuse, so the zygote has the full diploid number; **(1)** if meiosis didn't happen before fertilisation, the number of sets of chromosomes in the zygote would double each time. **(1)** *Another possible answer is that mixing of genes is important for variety in offspring.*

10. Clones

1. An individual that is genetically identical to another individual. **(1)**
2. **(a)** All the clones will have the gene for producing human hormones in their milk; **(1)** this is because they will have the same genes as the GM parent goat including the human hormone gene; **(1)** in sexual reproduction there is a chance that the offspring may not inherit the gene and so not produce the hormone. **(1)**
   (b) Any suitable answer, such as could be very expensive because it takes many attempts to make the clones, or cloned goats may suffer more health problems than normal goats. **(1)**

11. Stem cells

1. A cell that can divide to produce many kinds of differentiated cell. **(1)**
2. Embryos have to be destroyed when the stem cells are removed; **(1)** some people think it is wrong to do this because they believe embryos have a right to life. **(1)**
3. **(a)** Either: they are easier to extract, or they can produce more different kinds of cell. **(1)**
   (b) Adult stem cells from the patient will be recognised **(1)** but embryonic stem cells won't and the body will reject them. **(1)**

12. Protein synthesis

1. transcription, **(1)** translation **(1)**
2. **(a)** mRNA copies the base order from one strand of the DNA. **(1)**
   (b) tRNA brings amino acids to the ribosome. **(1)**
3. The order of the bases on the DNA strand is the template for the order of the bases on the mRNA; **(1)** the base order on the mRNA strand determines which tRNA molecules join in and in which order, **(1)** and this fixes the order that the amino acids are joined in the polypeptide. **(1)**

13. Proteins and mutations

1. A change in the base sequence in DNA/a gene. **(1)**
2. They are made from different sequences of amino acids **(1)** and the order and number of the amino acids affects the shape of the protein. **(1)**
3. Keratin forms fibrous proteins that produce tough fingernails; **(1)** a mutation in the gene that codes for keratin could change the order of amino acids in the protein; **(1)** the change in amino acids could change the shape of the protein so that it is not as strong. **(1)**

14. Enzymes

1. Enzymes change the rate of reactions; **(1)** they are biological catalysts because they are found in living organisms. **(1)**
2. **(a)** Any one suitable reaction that happens inside cells, such as DNA replication, protein synthesis. **(1)**
   (b) Any one suitable reaction where enzymes are secreted outside cells, such as digestion in the alimentary canal. **(1)**
3. The active site is a specially shaped area of the enzyme that matches the shape of the substrate **(1)** and is where the substrate molecule is held during the reaction. **(1)**

15. Enzyme action

1. Just below 40°C, **(1)** which is when the rate of reaction/enzyme activity is fastest. **(1)**
2. The hypothesis shows how shape is important to how the enzyme works **(1)** so that only a substrate with the right shape **(1)** will fit into the active site. **(1)**

16 and 17. Biology extended writing 1 and 2

Answers can be found on page 116.

18. Aerobic respiration

1. glucose + oxygen → carbon dioxide + water **(1)** *Energy may be included in brackets, to show that it is not a chemical substance.*
2. Respiration releases energy **(1)**, which organisms need for life processes/growth, movement etc. **(1)**
3. The concentration of oxygen is higher in the blood than in the tissues **(1)** because oxygen in the cells has been used in respiration/the blood contains oxygen collected in the lungs. **(1)** So oxygen diffuses down its concentration gradient from the blood into the cells, where it can be used in respiration. **(1)**

19. Exercise

1. As exercise level increases/gets more intense **(1)** the heart rate increases. **(1)** OR As exercise level reduces/gets easier **(1)** the heart rate decreases. **(1)** *It's not enough to say 'heart rate increases'. You need to link how heart rate changes as exercise level changes to get both marks.*
2. Any two from: Muscle cells are respiring faster because they are contracting faster. **(1)** Faster respiration needs more oxygen. **(1)** Faster breathing helps get oxygen into the blood more quickly. **(1)**
3. Cardiac output is heart rate multiplied by stroke volume. **(1)** A trained athlete will have a similar cardiac output because their stroke volume is much larger. **(1)**

20. Anaerobic respiration

1. The extra oxygen needed after exercise **(1)** that helps to break down lactic acid from anaerobic respiration. **(1)**
2. Running fast for 10 minutes will need more anaerobic respiration than running fast for 3 minutes. **(1)** More anaerobic respiration will produce more lactic acid, **(1)** which will need more oxygen to break it down. **(1)**
3. Anaerobic respiration produces lactic acid, **(1)** which isn't broken down until after exercise is completed. **(1)** If they used anaerobic respiration over such a long period, the concentration of lactic acid in their blood would become very high. **(1)** *Also less energy is released per glucose molecule by anaerobic respiration so the athlete would quickly run out of energy if they used more anaerobic respiration.*

21. Photosynthesis

1. reactants: water + carbon dioxide; **(1)** products: oxygen + glucose **(1)**
2. Only some leaf cells contain chloroplasts which is where photosynthesis occurs. **(1)**
3. Photosynthesis only occurs when there is enough light/during the day; **(1)** plants need carbon dioxide for photosynthesis; **(1)** carbon dioxide diffuses into the leaf through the stomata so the stoma are open during the day. **(1)** *The question doesn't ask why stomata close at night, so focus only on what happens during the day.*

22. Limiting factors

1. Something that limits the rate of photosynthesis **(1)** when it is at a low level. **(1)**
2. Photosynthesis produces oxygen. **(1)** The more rapidly oxygen is released, the faster the photosynthesis reactions must be happening. **(1)**
3. Shape of graph should be similar to the graph for light. **(1)** Labels should explain that temperature is the limiting factor as the graph slopes up **(1)** but that some other factor is limiting where the graph flattens off. **(1)** *Higher temperatures may also directly affect the enzymes that control the reactions in photosynthesis, and cause denaturation.*

23. Water transport

1. Water enters plant root cells by osmosis. **(1)** Dissolved mineral salts enter plant root cells by active transport. **(1)**
2. Xylem vessels carry water and dissolved mineral salts from the roots to all parts of the plant. **(1)** Phloem vessels carry sucrose around the plant to where it is needed or where it will be stored. **(1)**
3. Water is lost from the leaf to the air by transpiration through stomata. **(1)** This pulls water up the xylem through the plant. **(1)** This pulls water from the roots, **(1)** which causes water to move into the root from soil water. **(1)**

24. Osmosis

1. Net movement of water **(1)** from an area of high water concentration **(1)** to an area of low water concentration **(1)** through a partially permeable membrane. **(1)**
2. There is a higher concentration of water molecules on the right of the membrane than on the left. **(1)** Water molecules are small enough to pass through the partially permeable membrane so osmosis will occur from the right side to the left side of the membrane. **(1)** The volume on the left will increase as more water molecules cross the membrane, and volume on the right will fall. **(1)**
3. The potato strips would lose water by osmosis **(1)** because the solution is more concentrated than the cytoplasm of the potato cells. **(1)** So their weight would decrease. **(1)**

25. Organisms and the environment

1. The organisms may not be evenly spread, **(1)** so averaging several results helps to even out random variation. **(1)**

2. Count the number of snails in several randomly placed quadrats; **(1)** average the results and use this to calculate the population size in the whole grassland. **(1)**
3. average of samples is (3 + 1 + 2 + 0 + 5)/5 **(1)** = 11/5 = 2.2; **(1)** total area is 20 × 15 = 300 m^2; **(1)** area of one quadrat is 0.25 m^2, so estimated population size of snails is 2.2 × (300/0.25) **(1)** = 2640 **(1)** *Remember to show your working.*

26. Biology extended writing 3

Answers can be found on page 116.

27. Fossils and evolution

1. Fossils do not always form because conditions are not always suitable; **(1)** soft tissue decays so soft-bodied organisms rarely make fossils; **(1)** many fossils are too deep in the ground to have been discovered. **(1)**
2. Birds are not obviously related to any other group living at the moment (because of the lack of feathers elsewhere). **(1)** A feathered dinosaur is evidence that birds evolved from dinosaurs. **(1)**

28. Growth

1. Any one from: measure its change in height over time, measure its change in mass over time, measure the change in number and size of leaves over time (or similar). **(1)** *The key point to mention is change over time.* If the measurements increase over time, then the plant is growing. **(1)**
2. A percentile chart compares the growth (or an aspect of growth e.g. length, body mass) of an individual **(1)** against the growth (of the same aspect) of other individuals of the same sex and age. **(1)**
3. The increase in size of a balloon is not a permanent increase in size, **(1)** while the increase in size of a child is a permanent increase in the number of cells in the child's body. **(1)**

29. Growth of plants and animals

1. They can both develop into any kind of specialised cell in the organism. **(1)**
2. Plant growth happens because of cell division at the tip of the shoot **(1)** and cell elongation further away from the shoot or root tip. **(1)** Plant development happens as a result of differentiation of cells into specialised cells (such as xylem and phloem). **(1)**
3. Animal growth occurs as a result of cell division of stem cells in the embryo, baby and child. **(1)** Animal development is the result of cell differentiation of the stem cells into specialised cells (such as muscle or nerve cells). **(1)**

30. Blood

1. oxygen in red blood cells **(1)**; glucose in plasma **(1)** *You must say which substance is carried by each part of the blood. In an exam, you would not get marks for just saying 'red blood cells and plasma'.*
2. Some white blood cells surround and destroy pathogens. **(1)** Other white blood cells produce antibodies that destroy pathogens. **(1)**
3. Platelets respond to a break/wound in a blood vessel by triggering the blood clotting process. **(1)** The clot blocks the wound. **(1)** This prevents pathogens getting into the body. **(1)**

31. The heart

1. tissue: muscle **(1)** *other possible answers include tendons, surface linings, nervous tissue*; muscle cells **(1)** *other possible answers include surface lining cells, nerve cells/neurones, connective tissue*
2. **(a)** They prevent the blood flowing the wrong way through the heart when the heart muscles contract. **(1)**
   (b) Contraction of the left ventricle pumps blood all around the body, **(1)** which needs to pump with greater force to push blood a greater distance around the body. **(1)**
3. The blood that leaves the heart and goes around the body is fully oxygenated/contains more oxygen than if it was a mix of oxygenated and deoxygenated blood. **(1)** This means the body tissues get a greater supply of oxygen. **(1)**

32. The circulatory system

1. to transport materials around the body **(1)**
2. Arteries carry blood away from the heart; **(1)** veins carry blood towards the heart. **(1)**

33. The digestive system

1. digestion: in mouth, stomach and small intestine **(2)** (1 mark for 1 correct, 2 marks for 2 or more correct); absorption: small intestine **(1)**
2. The liver produces bile, which is released into the small intestine to help in the digestion of fats. **(1)** Food molecules absorbed from the small intestine into the blood are carried to the liver, where they are processed/changed into other molecules. **(1)**
3. Any four points from: It neutralises stomach acid, **(1)** which means that the enzymes of the small intestine can work at their optimum pH, **(1)** which means they digest food molecules faster. **(1)** Emulsifying fats **(1)** gives a larger area for lipases to work on so they can digest fat molecules faster. **(1)** Digesting food faster means that more small food molecules are available in the small intestine for absorption/more nutrients can be absorbed from the same amount of food. **(1)**

34. Villi

1. **(a)** large surface area due to large numbers of villi; **(1)** single thickness of cells that cover the surface of villi; **(1)** large capillary network inside each villus **(1)**
   (b) Large surface area increases the surface area for diffusion – this increases the rate of diffusion. **(1)** Single layer of cells means molecules do not have far to diffuse from inside the small intestine into the body, which increases the rate of diffusion. **(1)** Large capillary network carries absorbed food molecules away quickly, maintaining a steep concentration gradient, which maximises the rate of diffusion. **(1)**

35. Probiotics and prebiotics

1. a food that is supposed to make you healthier when you eat it **(1)**
2. They are supposed to encourage some kinds of bacteria in your digestive system. **(1)**
3. Scientific evidence is more likely to produce results that are reliable **(1)** because it comes from studies that have been carefully controlled to avoid bias. **(1)** *Also, if the studies are impartial (carried out by scientists who aren't working for the food company) then the evidence can usually be trusted more.*

36 and 37. Biology extended writing 4 and 5

Answers can be found on page 116.

Chemistry answers

38. Structure of the atom

1. copper has more than one isotope **(1)**
2. relative abundance of $^{123}_{51}Sb$ = 100% – 57% = 43% **(1)**; relative atomic mass = ((121 × 57) + (123 × 43))/100 = (6897 + 5289)/100 = 121.9 **(1)**

39. The modern periodic table

1. Any one from: P, S, Cl, Ar **(1)** *Give yourself the mark if you put Si. Silicon is actually a semi-metal, but you are not expected to know that for Additional Science.*
2. **(a)** He may have put iodine before tellurium **(1)** because iodine has a lower relative atomic mass than tellurium. **(1)**
   (b) He would have swapped them if he knew that iodine has similar properties to bromine and chlorine (or similar properties to the other elements in group 7), **(1)** and he knew that tellurium had similar properties to other elements in group 6. **(1)**

40. Electron shells

1. **(a)** 2.1 **(1)**
   (b) 2.6 **(1)**
   (c) 2.8.7 **(1)**

2. Any four from: It has 12 protons; **(1)** it has 12 electrons; **(1)** its electron arrangement is 2.8.2; **(1)** it has 3 occupied electron shells; **(1)** it is in period 3; **(1)** its outer shell has 2 electrons so it is in group 2; **(1)** if it is in group 2 it is a metal. **(1)**

41. Ions

1. **(a)** Ca^{2+} **(1)**
   (b) F^{-} **(1)**
   (c) K^{+} **(1)**
   (d) S^{2-} **(1)**
2.

lithium atom, Li
Li
F
fluorine atom, F
electron transferred
lithium ion, Li^{+}
Li
F
fluoride ion, F^{-}

(1 mark for drawing each atom and ion correctly; **1** mark for showing the electrons gained and lost correctly; **1** mark for labelling the atoms and ions correctly)

42. Ionic compounds

1. **(a)** CaS **(1)**
   (b) $Mg(NO_3)_2$ **(1)**
   (c) K_2SO_4 **(1)**
2. Beryllium has an atomic number of 4, so its electronic structure is 2.2 and it needs to lose 2 electrons to form an ion; **(1)** beryllium forms 2+ ions; **(1)** fluorine has an atomic number of 9, so its electronic structure is 2.7 and it needs to gain 1 electron to form an ion; **(1)** fluorine forms 1– ions. **(1)** The formula is BeF_2. **(1)**
   In an exam, if you worked out the charge on one of the ions incorrectly, but then used that charge in the correct way to work out a formula, you would still get the mark for the formula.

43. Properties of ionic compounds

1. **(a)** conducts, **(1)** because it is molten and the ions can move around **(1)**
   (b) does not conduct, **(1)** because it is solid and the ions cannot move **(1)**
   (c) conducts, **(1)** because it is dissolved/a solution and so the ions can move around **(1)**
2. **(a)** insoluble **(1)**
   (b) soluble **(1)**
   (c) insoluble **(1)**

44. Precipitates

1. **(a)** A precipitate will form **(1)** of calcium carbonate **(1)** because calcium carbonate is insoluble. **(1)**
   (b) No precipitate will form **(1)** because both of the other combinations of ions are soluble. **(1)**
2. any soluble salt of calcium (e.g. calcium nitrate, calcium chloride); **(1)** any soluble hydroxide (sodium hydroxide, potassium hydroxide, ammonium hydroxide) **(1)**

45. Ion tests

1. **(a)** calcium carbonate **(1)**
   (b) Calcium ions give a red flame; **(1)** substances containing carbonates give off carbon dioxide when acid is added to them; **(1)** this would produce the fizzing. **(1)**
   (c) Bubble the gas through limewater; **(1)** if the gas is carbon dioxide the limewater will turn milky. **(1)**

46. Chemistry extended writing 1

Answers can be found on page 116.

47. Covalent bonds

1. **(a)** 4 **(1)**
   (b) A hydrogen atom can only share/only needs to share one electron to fill its outer shell **(1)**; carbon needs to share four electrons **(1)** so it needs to share with four hydrogen atoms. **(1)**
   (c)

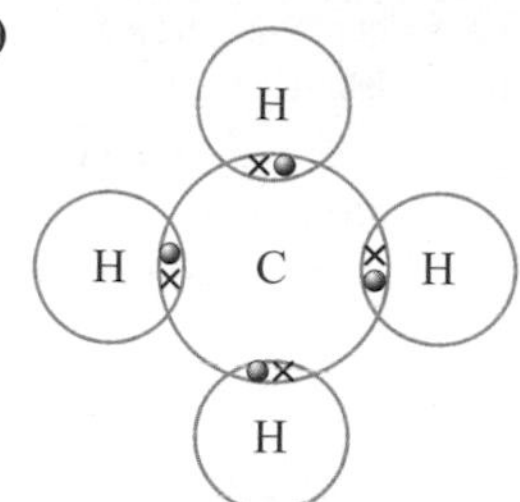

(**1** mark for showing the four hydrogen atoms surrounding one carbon atom, **1** mark for showing the dots and crosses properly). *You only need to show the outer shell of each atom.*

2. **(a)** Each oxygen molecule needs two electrons to complete its outer shell **(1)** so each atom must share two electrons. **(1)**
   (b)

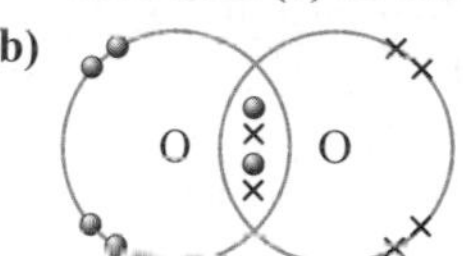

(**1** for showing two dots and two crosses where the electron shells overlap, **1** for correctly showing four other electrons in the outer shell of each oxygen atom.)

48. Covalent substances

1. **(a)** All the atoms are held together by strong bonds. **(1)**
   (b) There are no charged particles. **(1)**
   (c) It has a high melting point **(1)** because the atoms are all held together by strong bonds. **(1)**
2. Carbon dioxide forms individual molecules **(1)**, which are held together by weak forces **(1)**, so it has a much lower melting/boiling point than silicon dioxide. **(1)**

49. Miscible or immiscible?

1. Any two of the following points: Immiscible liquids will form two different layers **(1)** and if they were not mixed the person would only get some of the top layer on their salad **(1)**; shaking the bottle mixes the two layers up temporarily. **(1)**
2. **(a)** Nitrogen and oxygen need to be liquids as they go into the column **(1)** and this temperature is below the boiling points of both. **(1)**
   (b) This is below the boiling point of oxygen, so it stays a liquid/can be piped off **(1)**; but above the boiling point of nitrogen, so it evaporates/goes up the column. **(1)**

50. Chromatography

1. R_f = 4 cm/10 cm **(1)** = 0.4 **(1)**
2. the top spots for samples A, D and E **(1)**
3. Sample B: contains only one substance. **(1)** Sample C: two points from: contains 2 different substances; one is the same as sample B and part of sample E; one is the same as part of samples A and D (2 points needed for **1** mark). Sample D contains the same 3 substances as sample A. **(1)** Sample E: two points from: contains two substances; one is the same as a substance in A and D, one is the same as B and part of C (2 points needed for **1** mark).

51. Chemical classification 1

1. **(a)** simple molecular covalent substances **(1)**
   (b) ionic substances and giant molecular covalent substances **(1)**
   (c) giant molecular covalent substances **(1)**
2. Similarity: in both types of bonding the atoms end up with full outer electron shells; **(1)** both involve electrons in the outer shell. **(1)**
   Differences: in ionic bonding electrons are transferred to form ions but in covalent bonding the electrons are shared; **(1)** ionic bonding produces charged ions but covalent bonding produces neutral molecules; **(1)**

hemical classification 2

1. Metals conduct electricity when they are solids; **(1)** ionic substances only conduct electricity when they are molten or dissolved; **(1)** because the charged particles cannot move around when they are solid. **(1)** both involve electrons in the outer shell
2. **(a)** Some ionic or simple molecular covalent substances are soluble, **(1)** but no giant molecular covalent substances are soluble. **(1)**
   (b) Find out if the solution conducts electricity; **(1)** ionic substances conduct electricity when they are in solution (or simple molecular covalent substances do not conduct when they are in solution). **(1)**

53. Metals and bonding

1. **(a)** Metals consist of positive ions **(1)** in a regular arrangement, **(1)** surrounded by a sea of electrons/delocalised electrons. **(1)**
   (b) Some of the electrons from each atom can move between the ions, **(1)** and it is these electrons flowing in the same direction that form an electric current. **(1)**
2. **(a)** Malleable means the metal can be hammered into shape without breaking. **(1)**
   (b) When a force is applied to a metal **(1)** the layers of ions can slide over each other fairly easily. **(1)**

54. Alkali metals

1. $2Na(s) + 2H_2O(l) \rightarrow 2NaOH(aq) + H_2(g)$ (**1** mark for correct substances, **1** for state symbols, **1** for balancing)
2. Rubidium is more reactive than potassium; **(1)** any two from: the reactivity of the alkali metals increases as you go down the table, **(1)** because the outer electron in the atoms are further from the nucleus **(1)** and it is easier to remove them. **(1)**

55. Halogens

1. The reaction with chlorine will be faster **(1)** because chlorine is more reactive than bromine (or because the halogens get more reactive as you go up the group). **(1)**
2. sodium + bromine → sodium bromide; **(1)** $2Na + Br_2 \rightarrow 2NaBr$ (**1** mark for substances, **1** for balancing)

56. More halogen reactions

1. **(a)** Potassium bromide and chlorine will react **(1)** because chlorine is more reactive than bromine **(1)** so it will displace bromine from its compound. **(1)**
   (b) $2KBr + Cl_2 \rightarrow 2KCl + Br_2$ (**1** mark for correct reactants and products, **1** for balancing)
2. **(a)** $H_2(g) + Br_2(l)$ (**1** for reactants) $\rightarrow 2HBr(g)$ (**1** for products) (**1** mark for balancing, **1** for state symbols) *Don't forget that the gases you need to know about for this course form molecules with two atoms, so the formula for hydrogen gas is H_2, not H.*
   (b) dissolve it in water **(1)**

57. Noble gases

1. **(a)** density: any answer between about 2.3 and 2.8 **(1)** (although in an exam you would probably get the mark for anything above 2.15); boiling point: any answer between about −130 and −100°C **(1)** (in an exam you would probably get the mark for anything above −152).
   (b) The densities of the gases increase as you go down the table; **(1)** so xenon is likely to be denser than krypton. **(1)**
2. **(a)** Helium has a low density **(1)** so it makes the airship float. **(1)**
   (b) Argon is inert **(1)** so it does not react with the hot metal/stops oxygen coming into contact with the hot metal. **(1)**

58 and 59. Chemistry extended writing 2 and 3

Answers can be found on page 116.

60. Temperature changes

1. **(a)** It is an exothermic reaction; **(1)** combustion reactions release energy. **(1)**
   (b)

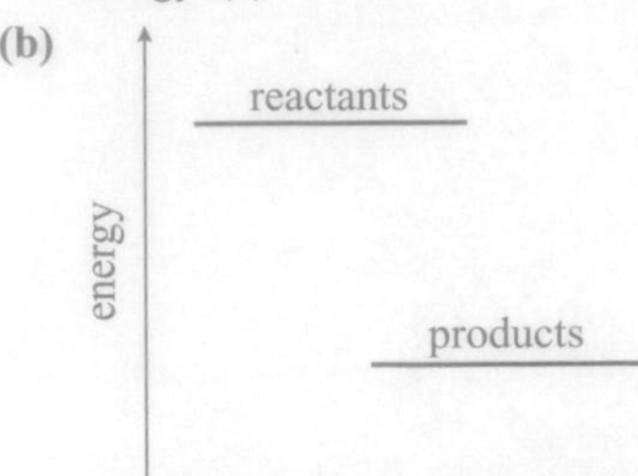

(**1** mark for labelling the energy axis, **1** for placing the reactants above the products)

2. When a reaction happens the bonds in the reactants have to be broken **(1)** and this takes in energy; **(1)** energy is released when the bonds in the products form; **(1)** for this reaction, the energy taken in to break the bonds is greater than the energy released when new bonds are formed. **(1)**

61. Rates of reaction 1

1. 1 kg is left; **(1)** catalysts do not get used up in reactions. **(1)**
2. **(a)** Rate of reaction will decrease **(1)** because there is a smaller surface area of the solid reactant **(1)** so there will be fewer collisions with the other reactant. **(1)**
   (b) Rate of reaction will decrease **(1)** because the particles will be moving more slowly **(1)** so there will be fewer collisions between reactants **(1)** and the collisions will be of lower energy and so the reaction may not happen when they collide. **(1)**
   (c) Rate of reaction will increase **(1)** because a given volume of a more concentrated solution contains more solute particles **(1)** so collisions between the reactant particles are more likely. **(1)**

62. Rates of reaction 2

1. Catalytic converters work best when they are hot; **(1)** the catalytic converter will be cold at the start of the journey **(1)** and may not convert all the carbon monoxide into carbon dioxide. **(1)**
2. Your graph should look like one of these.

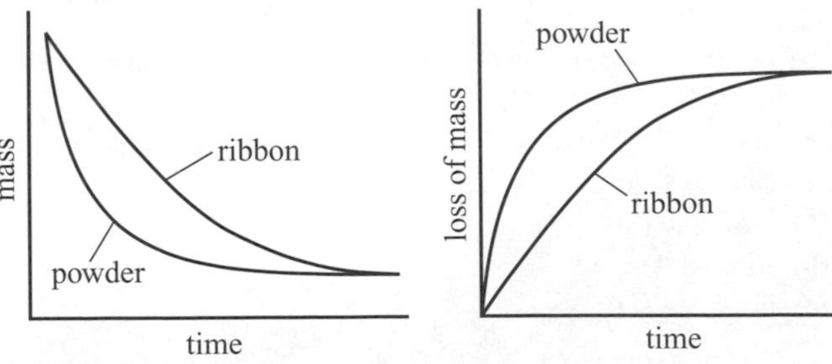

(**1** mark for showing the powder reaction losing mass faster than the ribbon reaction; **1** mark for showing the mass of both reactions the same at the end.) *Remember that changing the rate of a reaction does not change the total amount of product formed, only* how fast *it is formed.*

63. Relative masses and formulae

1. **(a)** $CaO = 1 \times Ca + 1 \times O$
   $M_r = 40 + 16 = 56$ **(1)**
   (b) $NH_3 = 1 \times N + 3 \times H$
   $M_r = 14 + (3 \times 1)$ **(1)**
   $= 17$ **(1)**
   (c) $CaCl_2 = 1 \times Ca + 2 \times Cl$
   $M_r = 40 + (2 \times 35.5)$ **(1)**
   $= 111$ **(1)**
2. **(a)** H_2O_2 **(1)**
   (b) HO **(1)**

64. Formulae and composition

1. Na: 8/23 = 0.348, O: 2.78/16 = 0.174 **(1)**
   Na: 0.348/0.174 = 2, O: 0.174/0.174 = 1 **(1)**
   Na_2O **(1)**
2. 1 atom of carbon, A_r for C = 12
   M_r = 12 + 4 = 16 **(1)**
   percentage by mass = 1 × 12/16 × 100 **(1)**
   = 75% **(1)**
3. 2 atoms of carbon, A_r for C = 12
   M_r = (2 × 12) + (6 × 1) + 16 = 46 **(1)** *Notice that there are hydrogen atoms at two different places in the formula.*
   percentage by mass = 2 × 12/46 × 100 **(1)**
   = 52% **(1)**

65. Masses of reactants and products

1. RFM CH_4: 12 + (4 × 1) = 16 **(1)**
   RFM O_2: 2 × 16 = 32 **(1)**
   16 g methane reacts with 32 g oxygen
   so 1 g methane reacts with 32/16 = 2 g oxygen **(1)**
   10 g methane reacts with 20 g oxygen **(1)**
2. $2K + Cl_2 \rightarrow 2KCl$ **(1)**
   RFM K: 2 × 39 = 78 **(1)**
   RFM KCl: 2 × (39 + 35.5) = 149 **(1)**
   78 g K produces 149 g KCl
   78/149 g K produces 1 g KCl
   20 x 78/149 g = 10.4 g K needed to produce 20 g KCl **(1)**

66. Yields

1. Theoretical yield is the amount of product that should be obtained worked out from the balanced equation; **(1)** actual yield is what is really obtained. **(1)**
2. percentage yield = 240 tonnes/600 tonnes × 100 **(1)** = 40% **(1)**

67. Waste and profit

1. The slag is useful, so the company can sell it to people who want to use it to make cement; **(1)** if they can sell it, it also means that they do not have to pay to dump it in landfill sites; **(1)** they make more profit. **(1)**

68 and 69. Chemistry extended writing 4 and 5

Answers can be found on page 116.

Physics answers

70. Static electricity

1. **(a)** Electrons have been transferred **(1)** from the comb to the cloth; **(1)** the comb now has fewer electrons than protons **(1)** so it has a positive charge. **(1)**
   (b) The comb will repel the rod (or the rod will move away) **(1)** because they both have the same charge (or because both charges are positive). **(1)**
2. The positive charge on the comb attracts electrons in the pieces of paper **(1)** so they move towards the surface of the paper closest to the comb; **(1)** this negative charge is attracted to the opposite charge on the comb **(1)** and the pieces of paper move towards the comb. **(1)**

71. Uses and dangers

1. You have built up a charge of static electricity as you walk across the carpet **(1)** because electrons have been transferred from the carpet onto you (or from you to the carpet), **(1)** and these electrons flow to earth (or from earth to you) when you touch the door, **(1)** which may give you a shock. **(1)**
2. The paint is given a charge of static electricity; **(1)** the droplets spread out because they all have the same charge; **(1)** the object being painted is given the opposite charge **(1)** so the paint is attracted to it. **(1)**

72. Electric currents

1. I = Q/t = 500 C/10 s **(1)**
   = 50 A **(1)**
2. t = Q/I **(1)** = 100 C/5 A **(1)** = 20 s **(1)**

73. Current and voltage

1.

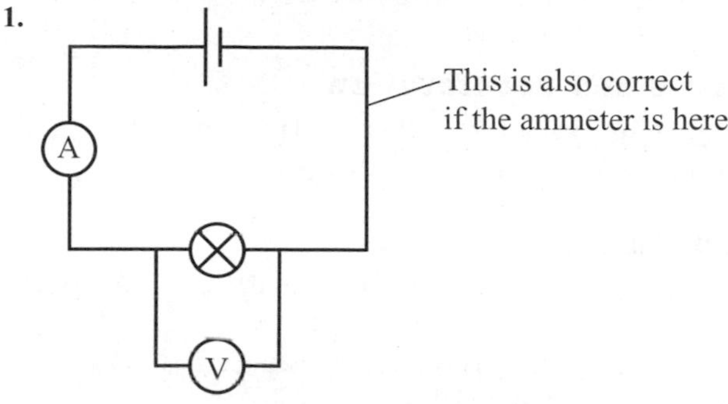

This is also correct if the ammeter is here.
A
V

(**1** mark if you have the symbols correct for the ammeter, voltmeter, cell and bulb; **1** mark for a drawing with no gaps in the circuit; **1** mark for the ammeter in series in the circuit and the voltmeter in parallel to the bulb)

2. **(a)** 3 A **(1)**
   (b) 1.8 A **(1)**
3. 5 V × 20 C **(1)** = 100 J **(1)**

74. Resistance, current and voltage

1. I=V/R **(1)** = 30 V/15 Ω **(1)** = 2 A **(1)**
2. R = V/I **(1)** = 6 V/2 A **(1)** = 3 Ω **(1)**

75. Changing resistances

1. **(a)** The current will increase **(1)** because the resistance of an LDR decreases when the brightness of light increases. **(1)**
   (b) The current will decrease **(1)** because the resistance of a thermistor increases when it is colder. **(1)**
   (c) The current will be zero **(1)** because a diode only conducts in one direction. **(1)** *You get no marks if you just say that the current decreases.*

76. Transferring energy

1. 10 minutes = 600 seconds **(1)**
   E = 9 A × 230 V × 600 s **(1)**
   = 1 242 000 J (**1** mark for correct answer, **1** for correct units)
2. I = P/V **(1)** = 3 W/6 V **(1)** = 0.5 A **(1)**

77. Physics extended writing 1

Answers can be found on page 117.

78. Vectors and velocity

1. distance = 80 m − 0 m = 80 m; **(1)** time = 60 − 0 = 60 s; **(1)** speed = 80 m/60 s; **(1)** = 1.33 m/s **(1)**
2. Acceleration is a vector quantity **(1)** so it has a direction as well as a size; **(1)** the car is accelerating because it is changing direction all the time it is travelling around the roundabout. **(1)**

79. Velocity and acceleration

1. initial velocity = 20 m/s, final velocity = 10 m/s, change in velocity = 10 − 20 = −10 m/s; **(1)** time taken = 20 s; acceleration = −10 m/s/20 s **(1)** = $-0.5\ m/s^2$ **(1)**
2. area of triangle between 70 and 80 seconds = $\frac{1}{2}$ × 10 s × 30 m/s = 150 m; **(1)** area of square beneath the triangle = 10 s × 10 m/s = 100 m; **(1)** area of rectangle between 80 and 90 seconds = 10 s × 40 m/s = 400 m; **(1)** distance travelled = 150 + 100 + 400m = 650 m **(1)**

80. Resultant forces

1. **(a)** A diagram like the one for the man standing on the floor, with labelled up and down arrows **(1)** and both arrows the same size. **(1)**

(b) her weight pushing down on the chair; **(1)** the chair pushing up on her **(1)**

2. (a) Its weight will be less **(1)** so now there will be an upwards resultant force on it. **(1)**

   (b) The change of weight does not affect the forward or rear forces **(1)** so the horizontal resultant force will remain zero. **(1)**

81. Forces and acceleration

1. $F = 0.45 \times 10^{-3}$ kg $\times 1.32 \times 10^3$ m/s^2 **(1)** = 0.594 N **(1)**
2. m = F/a **(1)** = 150 N/5 m/s^2 **(1)** = 30 kg **(1)**

82. Terminal velocity

1. Mass = 35/1000 kg = 0.035 kg; **(1)** W = 0.035 kg × 10 N/kg **(1)** = 0.35 N **(1)**
2. (a) Acceleration is zero **(1)** because she is at terminal velocity/is not getting any faster. **(1)**

   (b) 700 N **(1)** because if she is at terminal velocity/not accelerating her air resistance must be the same size as her weight. **(1)**
3. If her mass is greater her weight is greater, **(1)** so she will be falling faster (or it will take longer) **(1)** by the time the air resistance balances her weight. **(1)**

83. Stopping distances

1. (a) If the driver is tired their reactions will be slower **(1)** so they will travel further while they are reacting to the danger; **(1)** but being tired does not affect how well the car slows down once the brakes are applied. **(1)**

   (b) The car is moving normally/has not started to slow down during the thinking distance, **(1)** so the state of the road has no effect on this distance. **(1)**

   (c) The thinking distance is how far a car travels while the driver reacts to a danger, **(1)** so if the car is travelling faster it will move further during this time; **(1)** if it is travelling faster, the force from the brakes will take longer to stop it, **(1)** and it will travel further while it is coming to a stop. **(1)**

84. Momentum

1. momentum = 1200 kg × 15 m/s **(1)** = 18 000 **(1)** kg m/s **(1)**
2. Any three from: The total momentum before and after a collision is the same (or momentum is conserved); **(1)** momentum depends on velocity and mass; **(1)** the total mass of the car and van is greater than the mass of the van (or the mass of the moving objects has increased); **(1)** so the velocity must decrease. **(1)**
3. velocity = momentum/mass **(1)** = 300 kg m/s / 50 kg **(1)** = 6 m/s **(1)**

85. Momentum and safety

1. Any three from: it stops your head hitting the dashboard; **(1)** it is stretchy so you take a longer time to stop **(1)** than the rest of the car; **(1)** so the forces on you are smaller. **(1)**
2. It takes longer for the main part of your body to stop if you bend your knees **(1)** so the rate of change of momentum is less **(1)** and the forces on you are smaller. **(1)**
3. Any three from: A seatbelt reduces the forces on the passenger in a collision (or other explanation of the benefits of seatbelts) **(1)** so they reduce injuries; **(1)** if the collision is at high speed the seat belt may cause an injury, **(1)** but without this the injuries caused by the crash would be even greater. **(1)**

86. Work and power

1. 10 minutes = 10 × 60 seconds = 600 seconds **(1)**
   power = 60 000 J/600 s **(1)** = 100 W **(1)**
2. work done = 10 N × 1.5 m **(1)** = 15 J **(1)**
   power = 15 J/2 s; **(1)** = 7.5 W **(1)**

87. Potential and kinetic energy

1. (a) GPE = 0.5 kg × 10 N/kg × 2 m **(1)** = 10 **(1)** J **(1)**

   (b) v^2 = KE/(0.5 × m) **(1)** = 10 J/(0.5 × 0.5 kg) **(1)** = 40 (m/s)^2
   $v = \sqrt{40}$ = 6.32 m/s **(1)**

88. Braking and energy calculations

1. change in momentum = force × time = 16 N × 0.5 s **(1)** = 8 kg m/s; **(1)** velocity = momentum/mass = 8 kg m/s / 0.4 kg **(1)** = 20 N **(1)**
2. The kinetic energy of a moving object depends on velocity squared (or similar statement) **(1)**; 45 m/s is 3 times 15 m/s **(1)** so the braking distance will increase by 3^2 = 9 times, **(1)** so it will be 25m × 9 = 225 m. **(1)**

89 and 90. Physics extended writing 2 and 3

Answers can be found on page 117.

91. Isotopes

1. $^{11}_{5}B$
2. They both have 7 protons in the nucleus; **(1)** nitrogen-14 has 7 neutrons **(1)** and nitrogen-15 has 8 neutrons. **(1)**

92. Ionising radiation

1. Any four from: alpha particles have more mass than beta particles; **(1)** alpha particles have a positive charge but beta particles have a negative charge; **(1)** alpha particles are more ionising than beta particles; **(1)** alpha particles are less penetrating than beta particles; **(1)** alpha particles are helium nuclei and beta particles are electrons. **(1)**
2. Any three from: A beta particle has less mass than an alpha particle **(1)** so it transfers less energy when it hits something; **(1)** this makes it less likely to ionise particles it hits **(1)** but also means that it can travel further. **(1)**

93. Nuclear reactions

1. A fission reaction releases neutrons; **(1)** each neutron can make another nucleus split up **(1)**, which releases even more neutrons. **(1)**
2. Any four from: In a chain reaction each of the neutrons produced by fission can cause another nucleus to undergo fission **(1)** so the number of atoms undergoing fission increases very rapidly **(1)** and can cause an explosion; **(1)** in a controlled chain reaction some of the neutrons are absorbed by a different material **(1)** so only one neutron from each fission can cause another atom to fission. **(1)**

94. Nuclear power

1. The moderator slows down the neutrons **(1)** produced by fission reactions **(1)** to make them more likely to cause another fission reaction. **(1)**
2. Any four from: control rods absorb some of the neutrons **(1)** produced by the fission of uranium nuclei **(1)** so these neutrons cannot cause other uranium nuclei to fission; **(1)** if the control rods are pulled out of the core they will absorb fewer neutrons **(1)** so more fusion reactions will happen **(1)** and more heat energy will be released. **(1)**

95. Fusion – our future?

1. Fission is a large nucleus splitting up; **(1)** fusion is two small nuclei joining up. **(1)**
2. It needs to be published in a peer-reviewed journal, **(1)** and other scientists need to get the same results. **(1)**
3. Fusion reactions only happen at very high temperatures and pressures, **(1)** and currently more energy is needed to make the fusion reaction happen than it releases. **(1)**

96. Changing ideas

1. They could use tongs to handle the source **(1)** as this will keep their hands as far away from the source as possible; **(1)** they should make sure that the sources are never pointing towards people **(1)** to make it less likely that radiation will reach other people in the area. **(1)** *Teachers do not normally wear overalls or breathing apparatus, although these are precautions that radiation workers could take.*
2. The beta source is the most dangerous **(1)** as beta particles can penetrate several metres of air, **(1)** but alpha particles will be stopped before they get to the person/can only penetrate a few centimetres of air. **(1)**

97. Nuclear waste

1. Energy is needed to build **(1)** and decommission **(1)** nuclear power stations, so unless all this energy comes from nuclear power stations (or renewable resources) some carbon dioxide will be added to the atmosphere from burning fossil fuels. **(1)**
2. It produces less radioactivity/is less dangerous, **(1)** so it does not need to be stored in concrete/glass/below ground. **(1)**
   160 → 80 → 40 → 20 → 10 → 5

98. Half-life

1. Half of 60 is 30, half of 30 is 15 **(1)**, so the 4 hours represents 2 half-lives; **(1)** the half-life is 2 hours. **(1)**
2. 15 minutes is 5 half-lives **(1)** so the activity will be 5 Bq. **(1)**
   160 → 80 → 40 → 20 → 10 → 5

99. Uses of radiation

1. **(a)** Gamma rays are used to detect cancer by injecting a radioactive tracer into a patient; **(1)** a gamma camera detects where this collects, **(1)** and this shows where there is more cell activity than usual. **(1)**
   (b) Gamma rays are targeted at cancers/tumours **(1)** to kill the cells. **(1)**
2. Radon comes from rocks/ground/soil; **(1)** different types of rock are found in different parts of the country; **(1)** radon is a major contributor to the background so that higher levels of radon give a higher background count. **(1)**

100. More uses of radiation

1. If the paper is too thin more beta particles get through **(1)** and the pressure on the rollers will be reduced. **(1)**
2. Any four from: An alpha source inside the smoke alarm ionises the air, **(1)** which allows a current to flow between two plates; **(1)** smoke absorbs some of the alpha particles **(1)** so the current falls **(1)** and the alarm is sounded. **(1)**

101 and 102. Physics extended writing 4 and 5

Answers can be found on page 117.

Extended writing answers

Below you will find lists that will help you to check how well you have answered each Extended Writing question. A full answer will contain most of the points listed but does not have to include all of them, and may include other valid statements. Your actual answer should be written in complete sentences. It should contain lots of detail and link the points into a logical order. You are more likely to be awarded a higher mark if you use correct scientific language and are careful with your spelling and grammar.

16. Biology extended writing 1

Mitosis happens in most body cells; one cell produces two identical cells; called daughter cells; each cell contains a pair of each chromosome; this is the diploid number of chromosomes; and the cell makes copies of them; when the cell divides, each daughter cell ends up with one copy of each chromosome; daughter cells are identical to the parent cells; mitosis is used to grow and repair cells; meiosis happens in gametes; sperm and egg cells; chromosomes are copied as in mitosis; but the cell splits in two, then in two again; there are four daughter cells; each with only one copy of each chromosome; the daughter cells are not identical; they contain different alleles to each other; and are haploid.

17. Biology extended writing 2

mRNA/messenger RNA; formed in the nucleus of the cell; in a process called transcription; one strand of the DNA acts as a template; and mRNA forms as complementary bases against the DNA template; mRNA does not contain thymine; instead uracil is the complementary base to adenine; mRNA is single-stranded; mRNA passes out of the nucleus; through nuclear pores; where it travels to the ribosome; translation happens here; ribosome 'reads' along the mRNA; each triplet attracts a complementary triplet; this triplet is a tRNA molecule; attached to an amino acid; the tRNA joins to the mRNA; releasing its amino acid; which joins to other amino acids; to make a polypeptide.

26. Biology extended writing 3

Roots; have root hair cells; which increase surface area of root; water enters roots through osmosis; which is the movement of water; from area of high water concentration to low water concentration; across a partially permeable membrane; xylem takes water to all parts of the plant; up the stem; water has to move up against gravity; this requires an upward force; provided by transpiration; which is the evaporation of water from the leaf; and by the loss of water through the stomata; these help 'pull' the water up the plant; in a constant stream; using capillary action; called the transpiration stream.

36. Biology extended writing 4

Fossils can show gradual changes in organisms; so we can see how species have adapted to the environment; in the process of evolution; evidence that mammals had a common ancestor; can be seen through the pentadactyl limb; which we can see in this fossil; many mammals have the same bone structure in the forearm but adapted in different ways; e.g. wings of bats/flippers of whales/arms of humans/feet of horses; and in this early mammal fossil.

37. Biology extended writing 5

Small intestine contains enzymes which break down molecules; which are often insoluble; for example amylase which breaks down carbohydrates into simple sugars; proteases such as pepsin; break down proteins into amino acids; which are smaller than starting materials; and are soluble; they are absorbed; through the millions of villi; which are folds in the surface of the intestine; very large surface area; and very thin membrane; often only one cell thick; meaning short distance; for molecules to diffuse; so fast rate of absorption; also small intestine very long; villi have capillaries inside; in a big network; to take away absorbed nutrients; in the blood; and maintain concentration gradient; for diffusion.

46. Chemistry extended writing 1

The electronic structure of magnesium is 2.8.2; it has two electrons in its outer shell; an atom of magnesium becomes an ion by losing the two outer electrons; the ions have a charge of 2+; they are cations; the electronic structure of chlorine is 2.8.7; an atom of chlorine has 7 electrons in its outer shell; it gains one electron to form an ion; the ion has a charge of 1−; it is an anion; cations and anions attract each other; because they have opposite electrical charges; to form magnesium chloride; there have to be the same total number of electrons lost by magnesium atoms as gained by chlorine atoms; so in magnesium chloride there are two chloride ions for every magnesium ion; the formula is $MgCl_2$.

58. Chemistry extended writing 2

Oil and water are immiscible; they form separate layers; they can be separated using a separating funnel; which allows the water/bottom layer to be run off from the mixture; the main gases in air are oxygen and nitrogen; these mix together easily; and are gases at room temperature; so apparatus such as a separating funnel cannot be used; they can be separated because they have different boiling points; they are separated by fractional distillation; air is cooled to below the boiling point of both gases; so it becomes a liquid; it is passed into a fractionating column; the temperature in the fractionating column is below the boiling point of oxygen at the bottom; so most of it stays liquid and can be piped out; the temperature at the bottom is above the boiling point of nitrogen, so it stays as a gas; the temperature is colder near the top; so any evaporated oxygen condenses; but nitrogen remains a gas; and is removed from the column at the top.

59. Chemistry extended writing 3

Carbon dioxide consists of molecules of one carbon atom joined to two oxygen atoms; it is a compound; the atoms are held together by covalent bonds; each bond is a double bond; where four electrons are shared between the two atoms; carbon gets a full outer shell by sharing four electrons; oxygen atoms get a full outer shell by sharing two electrons; the covalent bonds are strong; but the bonds between separate carbon dioxide molecules are weak; so carbon dioxide has low melting and boiling points; and is a gas at room temperature.

Diamond consists of carbon atoms; held together in a lattice structure; by covalent bonds; it is an element; each carbon atom is bonded to four others; each bond is a single bond; where two electrons are shared between two atoms; the bonds are very strong; so diamond has high melting and boiling points; so it is a solid at room temperature; and is very hard.

For both substances, there are no charged particles that can move around freely; so neither substance will conduct electricity.

68. Chemistry extended writing 4

Exothermic and endothermic reactions cause changes in temperature; the temperature of the mixture can be measured using a thermometer; an exothermic change causes the mixture to get hotter; an endothermic change causes the mixture to get colder; when an acid is neutralised; bonds in the reactants are broken; this absorbs/needs energy; and bonds are made when the reactants form; this releases energy; the amount of energy absorbed is not usually the same as the amount released; if it takes more energy to break the bonds than it does to make them, energy is absorbed overall/the temperature falls; and the reaction is endothermic; if the amount of energy absorbed in bond breaking is less than the energy released, energy is released overall/the temperature rises; and the reaction is exothermic (or you can explain this by saying that if the reactants have higher bond energies than the products energy is released and the reaction is exothermic; if the products have higher bond energies than the reactants energy is absorbed overall and the reaction is endothermic).

69. Chemistry extended writing 5

Substances react with each other when particles of the reactants collide; with enough energy to make them react; so the rate of a reaction depends on how often collisions happen; and how many collisions have enough energy.

If the reaction involves a solid and a liquid, breaking the solid into smaller pieces increases the rate of reaction; because the smaller pieces have a greater surface area; so there is more chance for particles of the other reactant to hit it. Increasing the concentration of a liquid reactant increases the rate of reaction; because there are more particles of the reactant in a given volume; so the particles are more likely to collide with the other reactant. Increasing the temperature increases the rate of reaction; because particles have more energy at higher temperatures; and move faster; so there are likely to be more frequent collisions; and the collisions are likely to have more energy; so more of the collisions will result in a reaction.

77. Physics extended writing 1

The voltmeters would record the voltage supplied by the power supply; this will change as the voltage on the power supply changes; the ammeters will record the current through the circuit; this depends on the resistance; the current is given by the equations I = V/R; for 2 V, the current through the 5 Ω resistor would be 0.4 A (or any other calculation to show the current); as the voltage across the resistor is increased the current increases; the current is proportional to the voltage; because the resistance of the resistor does not change; for a low voltage the current through the filament lamp will be the same as through the resistor; as the voltage across the lamp increases the current increases; but it is not proportional to voltage; at higher voltages the current through the lamp is less than through the resistor; because the resistance of the lamp increases as it gets hot; because the ions in the metal vibrate more; and the moving electrons are more likely to bump into them; and transfer energy as heat/thermal energy.

89. Physics extended writing 2

The thinking distance is how far the car travels while the driver is deciding to stop; the time it takes for the driver to think does not change as the speed of the car changes; so the distance is a constant time multiplied by the speed; which is why the thinking distance is proportional to the speed; the braking distance is how far the car travels while the car is braking; work is done by the brakes when the car is braking; the work done is the braking force multiplied by the braking distance; the brakes convert all the car's kinetic energy to thermal energy; so for a constant braking force the distance is proportional to the kinetic energy; kinetic energy depends on velocity squared; so braking distance is proportional to velocity squared.

90. Physics extended writing 3

The gravitational potential energy of the ball depends on its height above the floor; the higher the ball, the more GPE it has; as it falls the GPE is converted to kinetic energy; some of this energy is transferred to thermal energy when the ball bounces; as the ball rises again the KE is transferred to GPE; as some energy has been wasted as thermal energy, the next bounce will not be as high; the efficiency can be measured by bouncing the ball and measuring the height of each bounce; the efficiency is calculated by taking the height from which the ball was dropped as representing the total energy, and the height of the first bounce as the 'useful' energy (or the actual GPE can be calculated at the top of each bounce using the height, g and the mass of the ball); the height can be measured by holding a metre rule near the ball and judging the maximum height; the measurement needs to be repeated several times; and an average value worked out; as this way of measuring may not be very accurate; or the bouncing ball and metre rule could be filmed using a video camera; so the image can be frozen to read the height.

101. Physics extended writing 4

Radioactive decay is a random process; we cannot predict whether a particular atom will decay; throwing a coin has a random result of heads or tails; throwing a die has a random result of one of 6 numbers; if a large number of coins are tossed, you can say that all the heads (or tails) represent atoms that decay in a certain time; the remaining coins can all be tossed again to model how many atoms decay in the next time interval; if you throw a large number of dice, you can say that all the dice that come up with a particular number decay in a certain time; throwing dice will model an isotope with a longer half-life than using a coin; because there is a 1 in 6 chance of modelling a decay; compared with a 1 in 2 chance for the coin; dice can be used to model isotopes with different half-lives by saying that more than one number can represent isotopes that decay.

102. Physics extended writing 5

Injecting into tumour: actinium-225 would be best; because it emits alpha particles; which do not penetrate far; so they would not harm tissues beyond the tumour; alpha particles are very ionising; so they are likely to damage/kill the cancer cells; the half-life of 10 days means the treatment could continue for a few weeks; to make sure all the cancer cells are killed.

Using as a tracer: gamma sources are used as tracers; because they nearly all pass through body tissues; so they can be detected by a gamma camera; and are not likely to damage healthy cells; technetium-99m would be best; because the half-life is fairly short; so the patient will not have radioactive material inside them for too long.

Beaming gamma rays into patient: cobalt-60 is best; it will continue to emit gamma rays for a long time; because it has a long half-life; so the machine can be used on many patients without needing to replace the isotope.

Your own notes

Your own notes

Published by Pearson Education Limited, a company incorporated in England and Wales, having its registered office at Edinburgh Gate, Harlow, Essex, CM20 2JE. Registered company number: 872828

www.pearsonschoolsandfecolleges.co.uk

Edited by Judith Head and Florence Production Ltd
Typeset by Tech-Set Ltd, Gateshead
Cover illustration by Miriam Sturdee

First published 2012

16 15 14 13 12
10 9 8 7 6 5 4 3 2 1

British Library Cataloguing in Publication Data
A catalogue record for this book is available from the British Library

ISBN 978 1 446 90265 3

Printed in Slovakia by Neografia

A note from the publisher
In order to ensure that this resource offers high-quality support for the associated Edexcel qualification, it has been through a review process by the awarding organisation to confirm that it fully covers the teaching and learning content of the specification or part of a specification at which it is aimed, and demonstrates an appropriate balance between the development of subject skills, knowledge and understanding, in addition to preparation for assessment.

While the publishers have made every attempt to ensure that advice on the qualification and its assessment is accurate, the official specification and associated assessment guidance materials are the only authoritative source of information and should always be referred to for definitive guidance.

No material from an endorsed revision guide will be used verbatim in any assessment set by Edexcel.

Endorsement of a revision guide does not mean that the revision guide is required to achieve this Edexcel qualification, nor does it mean that it is the only suitable material available to support the qualification, and any resource lists produced by the awarding organisation shall include this and other appropriate resources.